MW00760481

NanoScience and Technology

Springer
Berlin
Heidelberg
New York
Hong Kong
London
Milan
Paris
Tokyo

NanoScience and Technology

Series Editors: P. Avouris K. von Klitzing R. Wiesendanger

The series NanoScience and Technology is focused on the fascinating nano-world, mesoscopic physics, analysis with atomic resolution, nano and quantum-effect devices, nanomechanics and atomic-scale processes. All the basic aspects and technology-oriented developments in this emerging discipline are covered by comprehensive and timely books. The series constitutes a survey of the relevant special topics, which are presented by leading experts in the field. These books will appeal to researchers, engineers, and advanced students.

Sliding Friction
Physical Principles and Applications
By B. N. J. Persson

Scanning Probe Microscopy
Analytical Methods
Editor: R. Wiesendanger

Mesoscopic Physics and Electronics
Editors: T. Ando, Y. Arakawa, K. Furuya, S. Komiyama, and H. Nakashima

Biological Micro- and Nanotribology
Nature's Solutions
By M. Scherge and S. N. Gorb

Nanoelectrodynamics
Electrons and Electromagnetic Fields in Nanometer-Scale Structure
Editor: H. Nejo

Semiconductor Quantum Dots
Physics, spectroscopy and Applications
Editors: Y. Masumoto and T. Takagahara

Semiconductors Spintronics and Quantum Computation
Editors: D.D. Awschalom, D. Loss, and N. Samarth

Nano-Optoelectronics
Concepts, Physics and Devices
Editor: M. Grundmann

Noncontact Atomic Force Microscopy
Editors: S. Morita, R. Wiesendanger, and E. Meyer

Series homepage – http://www.springer.de/phys/books/nst/

Hitoshi Nejo (Ed.)

Nanoelectrodynamics

Electrons and Electromagnetic Fields in Nanometer-Scale Structure

With 126 Figures

Springer

Professor Hitoshi Nejo
National Research Institute for Metals
1-2 Sengen, Tsukuba 305-0047, Japan

Series Editors:
Professor Dr. Phaedon Avouris
IBM Research Division, Nanometer Scale Science & Technology
Thomas J. Watson Research Center, P.O. Box 218
Yorktown Heights, NY 10598, USA

Professor Dr., Dres. h. c. Klaus von Klitzing
Max-Planck-Institut für Festkörperforschung, Heisenbergstrasse 1
70569 Stuttgart, Germany

Professor Dr. Roland Wiesendanger
Institut für Angewandte Physik, Universität Hamburg, Jungiusstrasse 11
20355 Hamburg, Germany

ISSN 1434-4904
ISBN 3-540-42847-X Springer-Verlag Berlin Heidelberg New York

Library of Congress Cataloging-in-Publication Data applied for
A catalog record for this book is available from the Library of Congress.
Bibliographic information published by Die Deutsche Bibliothek
Die Deutsche Bibliothek lists this publication in the Deutsche Nationalbibliografie;
detailed bibliographic data is available in the Internet at http://dnb.ddb.de

Springer-Verlag Berlin Heidelberg New York
a member of BertelsmannSpringer Science+Business Media GmbH

http://www.springer.de

Typesetting: Camera-ready by the authors using a Springer TEX macropackage
Final Layout: EDV-Beratung Frank Herweg, Leutershausen
Cover design: *design & production*, Heidelberg

Printed on acid-free paper SPIN: 10758207 57/3141/tr - 5 4 3 2 1 0

Preface

Many books on mesoscopic systems have been published as progress has continued in the fields of nanoscience and nanotechnology. The focus in these books is mainly on quantum mechanical behavior in artificial electronic systems fabricated by nanometer-scale structuring. Such quantum mechanical behavior is projected to macroscopic observers and the quantum nature can be utilized in practical devices. Quantum computers, another hot topic nowadays, are characterized by excitation coherence properties among nanostructures, and the ability to maintain excitations is very important when using the characteristics as information. In that sense, the device is described as a microscopic system and some processes occur before being projected to macroscopic observers. In this book, the authors try to describe not only the techniques for fabricating nanostructures but also new directions as regards exciting systems and understanding how energy is dissipated through observation.

The idea of 'nano-electrodynamics' underlying the book is an analogy with the well-established classical electrodynamics. In contrast to the latter, 'nano-electrodynamics' is still in its infancy and far from well established. When a structure is miniaturized as a device, it is essential to have control over energy excitation and dissipation. Otherwise, when a device is squeezed down beyond a certain size and the energy dissipation becomes overwhelmed, the device will eventually collapse. It is our aim in this book to provide some thoughts on the task of making devices out of small structures.

The book is partly the result of the 1999 Tsukuba symposium on the interaction of electrons and electromagnetic fields. I wish to thank all the participants of the symposium for their contributions and fruitful discussion. I would especially like to acknowledge the great support from Dr. H. Hori, Dr. Z.-C. Dong, Dr. I.S. Osad'ko, Dr. A.G. Vitukhnovsky, Dr. J. Bae, Dr. J.-G. Hou, and Dr. N. Yamada. I also thank Dr. K. Amemiya for his help in compiling this book in its present form.

I would also like to thank Dr. Claus von Klitzing and Dr. Claus Ascheron for their original suggestion and encouragement in preparing this book. Finally I would like to thank Adelheid Duhm at Springer for checking all the manuscripts.

Tsukuba, Japan,
August 2002

Hitoshi Nejo

Contents

List of Contributors

Jongsuck Bae
Research Institute
of Communication
Tohoku University
2-1-1 Katahira, Aoba-ku
Sendai 980-8577, Japan
bae@riec.tohoku.ac.jp

Zhen-Chao Dong
National Institute
for Materials Science
1-2-1 Sengen, Tsukuba
Ibaraki 305-0047, Japan
DONG.Zhen-Chao@nims.go.jp

Daisuke Fujita
National Institute
for Materials Science
1-2-1 Sengen, Tsukuba
Ibaraki 305-0047, Japan
Fujita.Daisuke@nims.go.jp

Hirokazu Hori
Jamanashi University
4-3-11 Takeda
Kofu 400-8511, Japan
hirohori@es.yamanashi.ac.jp

J.G. Hou
Structure Research Laboratory
and Open Laboratory of Bond
Selective Chemistry
University of Science and Technology
of China
Hefei 230026
P. R. China
jghou@ustc.edu.cn

Ryo Ishikawa
Research Institute
of Communication
Tohoku University
2-1-1 Katahira, Aoba-ku
Sendai 980-8577, Japan
issi@riec.tohoku.ac.jp

Bin Li
Open Laboratory
of Bond Selective Chemistry
University of Science and Technology
of China
Hefei 230026
P. R. China
libin@mail.ustc.edu.cn

Shinro Mashiko
Communication Research
Laboratory
Kobe, Hyogo 651-2401, Japan
masahiko@crl.go.jp

Koji Mizuno
Research Institute
of Communication
Tohoku University
2-1-1 Katahira, Aoba-ku
Sendai 980-8577, Japan
koji@riec.tohoku.ac.jp

Hitoshi Nejo
National Institute
for Materials Science
1-2-1 Sengen, Tsukuba
Ibaraki 305-0047, Japan
NEJO.Hitoshi@nims.go.jp

Taizo Ohgi
National Institute
for Materials Science
1-2-1 Sengen, Tsukuba
Ibaraki 305-0047, Japan
OHGI.Taizo@nims.go.jp

Takayuki Okamoto
RIKEN (Institute of Physical
and Chemical Research)
Wako, Saitama 351-0198, Japan
okamoto@postman.riken.go.jp

I.S. Osad'ko
Lebedev Physical Institute RAS
Leninsky Prospect, 53
Moscow 119991, Russia
osadko@sci.lebedev.ru

Toshifumi Terui
Communication Research
Laboratory
Kobe, Hyogo 651-2401, Japan
terui@crl.go.jp

Alexei G. Vitukhnovsky
Lebedev Physical Institute RAS
Leninsky Prospect, 53
Moscow 119991, Russia
alexei@sci.lebedev.ru

Shiyoshi Yokoyama
Communication Research
Laboratory
Kobe, Hyogo 651-2401, Japan
syoko@crl.go.jp

Norifumi Yamada
Fukui University
3-9-1 Bunkyo
Fukui 910-8507, Japan
fyamada@ilnws1.fuis.fukui-u.ac.jp

1 Introduction: Electron and Photon Systems

H. Nejo and H. Hori

While previous books about nanometer scale phenomena have dealt with either electronic states or electromagnetic fields, this book is concerned with both as a unified whole. The characteristics of electrons or electromagnetic fields apparent at nanometer scales are quite different from those at the macroscopic scale. Nowadays these areas are called nanoscience or nanotechnology. They attract the interest of a wide range of people for their applications to new branches of technology [1–15].

When we deal with electromagnetic fields at the nanometer scale, the scale of interest is much shorter than the wavelength of light, at approximately 1 μm. The characteristics of such electromagnetic fields are therefore quite different from those pertaining to macroscopic sizes. The former are characterized as localized waves, while the latter are characterized as propagating waves. The former nanoscale wave is usually called an evanescent wave and its characteristics have now been studied by many researchers [16–50].

For electron waves, the situation is similar to that of a propagating light wave in that an electron can propagate through crystalline metals and semiconductors, whereas on nanoscales, when a metal block is cut in half, the propagating electron wave starts to penetrate into the vacuum from the surface. The nature of this penetrating wave can be very similar to that of an evanescent light wave. The electron wave penetrating the vacuum appears as a tunneling phenomenon: the electron can pass through a potential barrier. These features have been studied using the scanning tunneling microscope (STM) [51].

Consider the relationship between an electron and an electromagnetic field. Electromagnetism shows that the moving charged particle causes an energy flow, called a Poynting vector. Thus the field of the electron and the electromagnetic field are closely related. This interaction has been well studied at the macroscopic scale but not at the nanoscale.

One can see the interaction of an electron and the electromagnetic field in the case of a Coulomb blockade [52–62]. A Coulomb blockade can be described as follows. A double tunnel junction is biased by a battery. When the battery does work and the Fermi level of the emitter reaches the chemical potential of the central electrode of the double tunnel junction, electrons can tunnel through the first tunnel barrier by overcoming the Coulomb energy $e^2C/2$, where C is the macroscopic capacitance of the tunnel junction. When an electron tunnels through the potential barrier, the Poynting vector crosses a resistor that is connected to the double tunnel junction, indicating that current flows through the junction.

While the interaction between electrons and electromagnetic fields is a most interesting phenomenon, it has not yet been studied properly at the nanometer scale. The reason for this can be entirely attributed to the lack of technological development both in realizing nanometer scale structures and also in connecting them to excitation sources and detectors. For example, the interaction between a penetrating electron wave, i.e., a tunneling electron, and an electromagnetic field has been studied. A planar tunnel junction was prepared and then irradiated by light. Since a tunneling current was detected, the existence of a penetrating or decaying electron wave was deduced. However, as both the electrode and the tunnel barrier were irradiated, it was not possible to extract the net interaction effect between the tunneling electron and the evanescent field from the interaction between the propagating electron wave and the electromagnetic field in the electrodes. Due to the experimental set up, the details of the interaction were not revealed.

A similar situation occurred in an experiment involving the interaction between a propagating electron and an evanescent field, although in this case not involving a tunneling electron (Chap. 5). Light was introduced into an insulating slab through which the propagating electron flowed. Since the nature of the electromagnetic field in an insulator is evanescent, the propagating electron was assumed to interact with the evanescent wave in the slab. Light emission was observed after the electron reached a non-fluorescent screen. This is thought to be a consequence of the interaction between the propagating electron and the evanescent field. However, as the propagating electron interacts with the material of the slab itself, one cannot distinguish the interaction of the electron and the evanescent field from the interaction between the electron and the matter.

These examples show that it is no trivial task to set up an experiment that distinguishes the interaction between a tunneling electron and an evanescent field from the other possibilities. In this book, the reader will discover some efforts in this direction. Owing to the recent development of nanoscience and nanotechnology, we now possess a variety of tools for realizing experimental setups that can probe the interaction between a tunneling electron and an evanescent field.

The purpose of this book is to show the importance of constructing nanoscale electric and electromagnetic fields connected to a reservoir. We start with a conventional electric or electromagnetic field treatment, for which many theoretical and experimental results have been accumulated, and then move on to the importance of the reservoir connection.

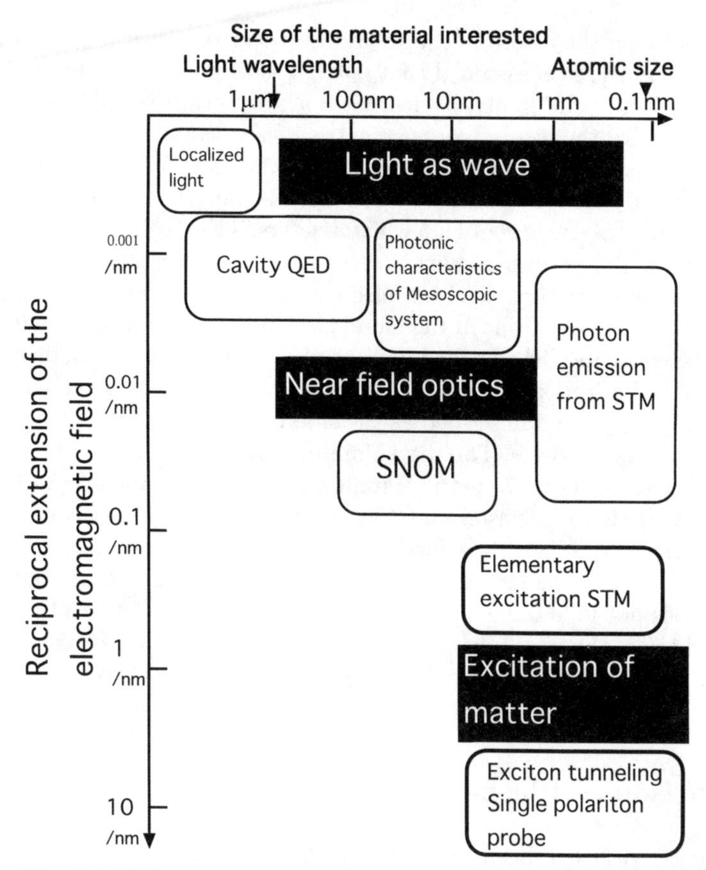

Fig. 1.1. Relation between the size of matter and the extension of the electromagnetic field that can be detected. Cavity QED is described in Sect. 1.2, SNOM in Sect. 1.3, photon emission from STM in Sect. 1.4, elementary excitation STM in Sect. 1.5, single polariton probes in Sect. 1.6 and mode connection to the macroscopic mode is described in Sect. 1.7

1.1 Size Considerations Concerning Electromagnetic Fields

Size considerations of electromagnetic fields are illustrated in Fig. 1.1 [63]. Cavity quantum electrodynamics (QED) is described in Sect. 1.2, scanning near-field optical microscopy (SNOM) in Sect. 1.3, photon emission from STM in Sect. 1.4, elementary excitation STM in Sect. 1.5, single polariton probes in Sect. 1.6 and mode connection to the macroscopic mode in Sect. 1.7.

Single-electron tunneling is accompanied by the creation of a quantum $\hbar\omega$ of electromagnetic field in accordance with the following energy balance: $eV = E_0 + \hbar\omega$. Whenever the microscopic electronic state changes, the energy changes by the unit quantum from the overall mode function. When the energy transfer between the electronic state and the electromagnetic state is described in this manner, it is assumed that one is treating a mode function whose size is three times the light wavelength. Such a picture is applicable to a laser cavity, where the size of confinement is of the same order as the wavelength. We are concerned with electromagnetic fields of much smaller size. When a molecule is placed under an STM tip, the gap size between the tip and the substrate is of the order of a few nanometers. The electromagnetic field in this nanometer-sized gap is much smaller than the conventional size described by the mode function. Such an electromagnetic field is described by the combination of the electrons in the metal electrode (tip) and the mode function on the larger scale, the so-called dressed photon. On an even smaller scale, the effect of radiation on a tunneling electron was studied by Tien and Gordon in the early 1960s and described as the side-band effect of the initial and final electronic states [64,65]. The total Hamiltonian H is then described by $H = H_0 + eV \cos \omega t$, where H_0 is the Hamiltonian excluding the effect of the electromagnetic field. Johannson and Xiao discussed ways of describing the electromagnetic field. There is a further complication for structures in an electromagnetic field at nanometer scales. A dressed photon has a large $\boldsymbol{k}$ vector which does not follow the dispersion relation $\omega/c = |\boldsymbol{k}|$. Due to this large $\boldsymbol{k}$ vector, there is the possibility of transition from the initial $\boldsymbol{k}$ vector $\boldsymbol{k}_\mathrm{i}$ to the final $\boldsymbol{k}$ vector $\boldsymbol{k}_\mathrm{f}$ via various states interacting with the $\boldsymbol{k}$ vector of the electron.

1.2 Comparison of Plasmon and Exciton

The field of cavity QED has been well studied [66–82]. The behavior of a quantum-well exciton resonantly coupled to a single-mode microcavity field in a linear (low-excitation) regime can be modeled in terms of coupled harmonic oscillators between which the energy is coherently recycling back and forth. When an exciton is coupled with a polariton, the state is described as an exciton–polariton. When a plasmon is coupled with a polariton, it is described as a plasmon–polariton. This is well defined when the size under consideration is at most half a wavelength. Below half a wavelength, the situation changes drastically. At such small scales, the mode is no longer defined. This is the situation when we try to see an elementary excitation or a single polariton. The difficulty can be brought out using STM as described in the next section. Electromagnetic fields in the nanometer region can be measured using light emission from a molecule under an STM tip excited by tunneling electrons. For any measurement a probe is required to detect signals. STM uses a probe that is very close to the sample. The tip–sample distance is usually only a few nanometers so that proximity effects always affect the electromagnetic field. Light emission from a molecule under an STM tip has been detected by two methods and the differences induced by different tip materials have been compared using tungsten and ITO-coated fiber tips (Chap. 2).

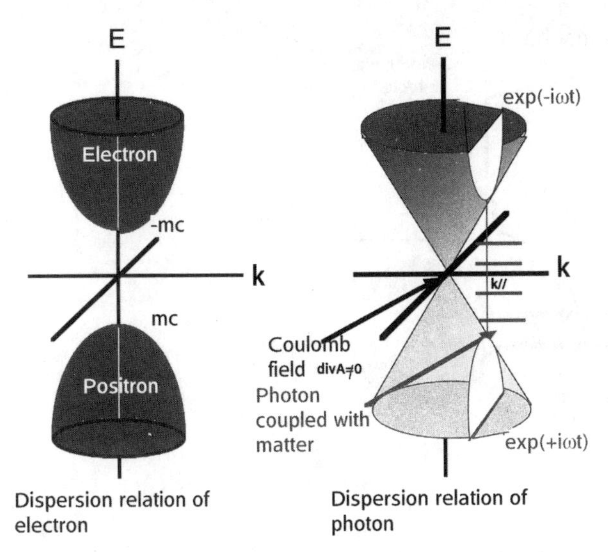

Fig. 1.2. Dispersion relations of an electron (*left*) and a photon (*right*). Photon states coupled with matter are denoted by *horizontal lines*. These have a larger $\boldsymbol{k}$ vector than those which obey the dispersion relation. SNOM detects these large $\boldsymbol{k}$ vector states

1.3 Coupling of Light with Plasmons

Figure 1.2 shows the dispersion relation of electrons and photons. SNOM uses this large $\boldsymbol{k}$ photon state. The effect between light and an insulator has been well studied. Ebbesen has also demonstrated an effect between light and metal. When a two-dimensional array on a metal film is irradiated with light, transmission efficiency can exceed unity when normalized to the area of the holes. This effect has also been studied by SNOM using a metallic tip or particle as scatterer [83–90].

1.4 Photon Emission from STM

Photon emission from an STM is explained by dipole moment radiation due to a tip-induced surface plasmon (Fig. 1.3) [91–105]. The electron–photon interaction can be considered by borrowing the idea of virtual photons. When an electron with $\boldsymbol{k}$ vector $\boldsymbol{k}_{\mathrm{e}}$ tunnels through a potential barrier, it can interact with the potential surface and change its state to $\boldsymbol{k}'_{\mathrm{e}}$. When the tunneling electron then experiences inelastic tunneling, a photon with $\boldsymbol{k}$ vector $\boldsymbol{k}'_{\mathrm{p}}$ is generated and the electron leaves the potential barrier. Since the photon is a virtual photon, it can interact with matter to borrow $\boldsymbol{k}$ vector and change its state to $\boldsymbol{k}$ vector $\boldsymbol{k}_{\mathrm{p}}$ (Fig. 1.4) [106].

This electron–photon interaction in a potential barrier in real space is illustrated in Fig. 1.5. When we prepare an electron path with a potential

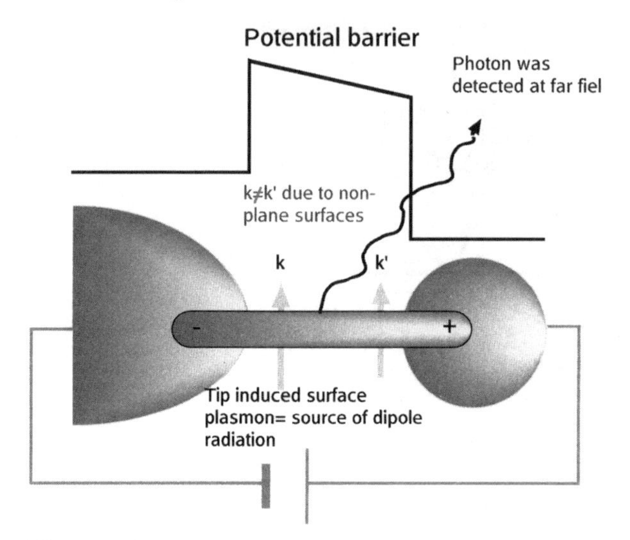

Fig. 1.3. Schematic diagram showing the tip-induced surface plasmon. When an STM tip is biased, a surface plasmon is induced between the STM tip and the metal surface. The wave vector $\boldsymbol{k}$ is not conserved, since the periodicity in the lateral direction is broken. This surface plasmon is the source of the dipole moment radiation

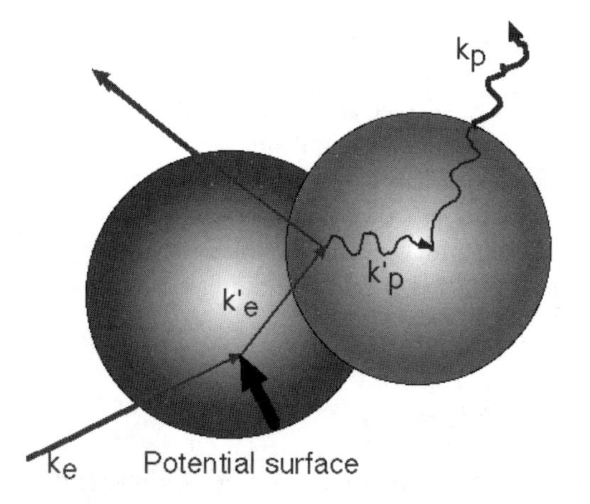

Fig. 1.4. Schematic diagram of the interaction between an electron and a photon on a potential surface. When an electron with $\boldsymbol{k}$ vector $\boldsymbol{k}_{\mathrm{e}}$ interacts with the potential, it may borrow virtual $\boldsymbol{k}$ to become $\boldsymbol{k}_{\mathrm{e}}'$. This electron interacts with another potential and emits a virtual photon with $\boldsymbol{k}$ vector $\boldsymbol{k}_{\mathrm{p}}'$. It then returns to a real electron. The virtual photon $\boldsymbol{k}_{\mathrm{p}}'$ interacts with the potential and recovers to a real photon $\boldsymbol{k}_{\mathrm{p}}$

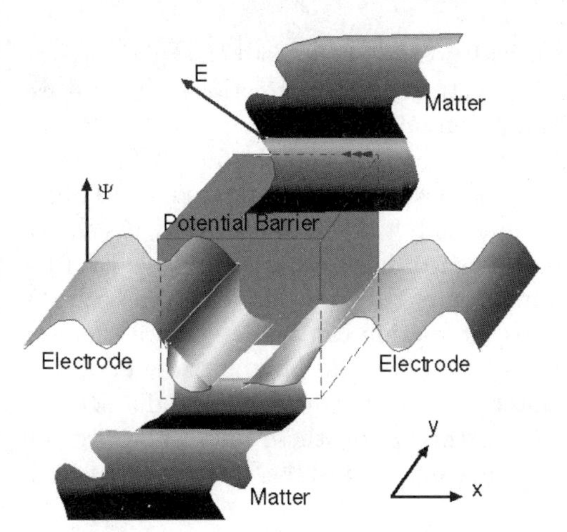

Fig. 1.5. Schematic diagram of the interaction between an electron and a photon in a potential barrier. When an electron Ψ tunnels through a potential barrier it may interact with a virtual photon which is the electromagnetic field E coupled with matter

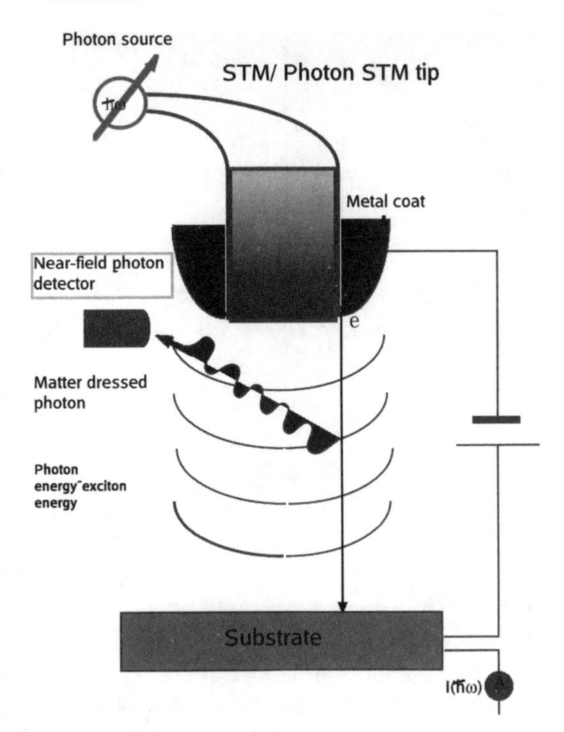

Fig. 1.6. Experimental setup showing the electron–photon interaction based on the possibility of interaction in a potential barrier. The evanescent field is introduced into the tunnel gap through which the electron tunnels. The electron–photon interaction is detected as a matter-dressed photon by a near-field photon detector

barrier in the middle (x direction), it is possible to introduce a path for the photon which is perpendicular to the electron path (y direction). The process described above can occur in such a potential barrier.

Photon-assisted tunneling was studied by Tien and Gordon in the 1960s. At that time they used a planar junction. Later they illuminated an STM tip but the state of the photon in the tunnel gap was not well defined. As shown in Fig. 1.6, by using a metal-coated fiber tip, if an evanescent wave is introduced into the tunnel gap, the interaction of the tunneling electron with the evanescent wave is well defined. This interaction can be detected as current $I(\omega)$ and as a matter-dressed photon by a near-field photon detector. Using this setup, the lateral size of the interaction is restricted by the diameter of the aperture of the fiber tip. Again, in this setup, the evanescent wave can excite the substrate surface and the net effect must be distinguished from this photon–substrate interaction.

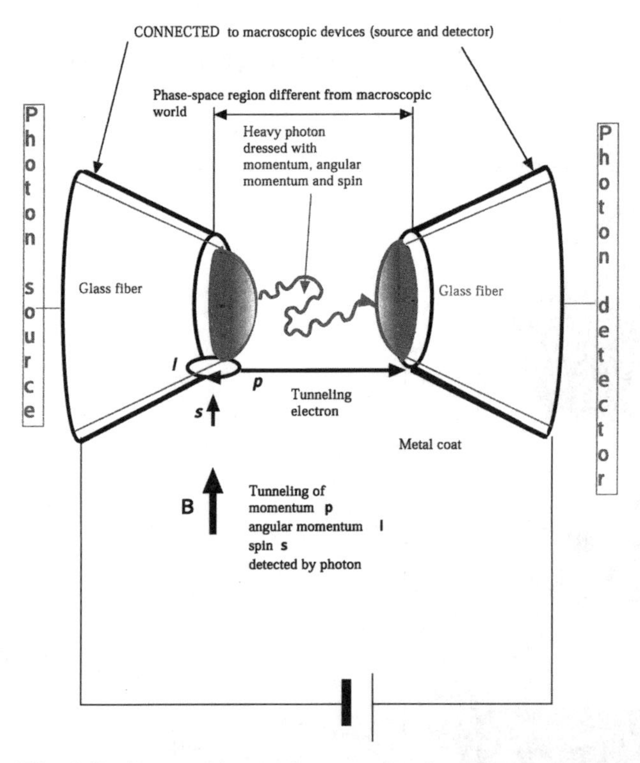

Fig. 1.7. An experimental setup showing the interaction of a tunneling electron with a photon. Two metal-coated fiber tips face each other. When they come close enough, an electron tunnels through the potential barrier. A dressed photon tunnels across the gap. The interaction of the electron states, such as momentum $\boldsymbol{p}$, angular momentum $\boldsymbol{I}$ and spin $\boldsymbol{s}$, with the photon can be detected by a photon detector

It should be possible to detect tunneling of momentum, angular momentum and electron spin using two metal-coated fiber tips (Fig. 1.7). When two such tips are placed face to face in close proximity, electrons can tunnel between the two metallic electrodes. If light is introduced at one fiber tip, an evanescent field is generated between the two fiber tips. This heavy photon, dressed with electron momentum, angular momentum and spin, can be detected by another fiber tip connected to a photon detector. Gimzewski et al. pioneered the study of photon emission from an STM tip and sample in 1988 [107–116]. When a bias voltage is applied across a tunneling barrier so that an electron tunnels, the electron in the counter electrode has an excess potential energy equal to eV where e is the electronic charge and V the bias voltage. The electron will relax by releasing this excess energy. Most of the excess energy is converted into heat by collision with phonons, but some fraction of the excess energy will be emitted as a photon under appropriate conditions. If the bias voltage is a few volts, the emitted photon has energy corresponding to the visible region. When one observes such photon emission from a pristine metal surface, the photon emission from an STM is considered as dipole moment radiation due to the tip-induced surface plasmon. In the surface plasmon view, photon emission from an STM can be described as follows [117–120]. When the tip–sample separation is no more than 0.5–1 nm, the surface plasmons on the two surfaces interact strongly and form an interface–plasmon mode. The resonance in the light emission spectrum occurs at the frequency for which half a wavelength of the interface mode fits into the cavity between tip and sample (although later, the reader will discover that in reality this does not apply to such a small system). The charge oscillations associated with the interface plasmon have opposite signs to the two electrodes. Photon emission from an STM as described via the surface-plasmon–polariton mode will be described in the next section.

1.5 Photon Emission from a Molecule Under an STM Tip

To observe photon emission, the object of interest should be connected to some macroscopic mode [121–131]. Here the object of interest is a single molecule, a typical quantum mechanical object [132]. When the molecule is excited by a tunneling electron, the dynamic dipole moment couples with the surface plasmon between the tip and the substrate. This surface plasmon is localized between the tip apex and the counter substrate surface. It is a coupled mode of the electromagnetic field in the cavity (formed by the tip and the substrate surface) and the electrical system of the tip and the substrate (and the molecule itself) [133,134]. This surface plasmon propagates along the surface of the tip and/or the substrate surface and finally connects with the macroscopic mode. The size of the SNOM electromagnetic field is a few tens of nanometers between the tip and sample, since the tip height is

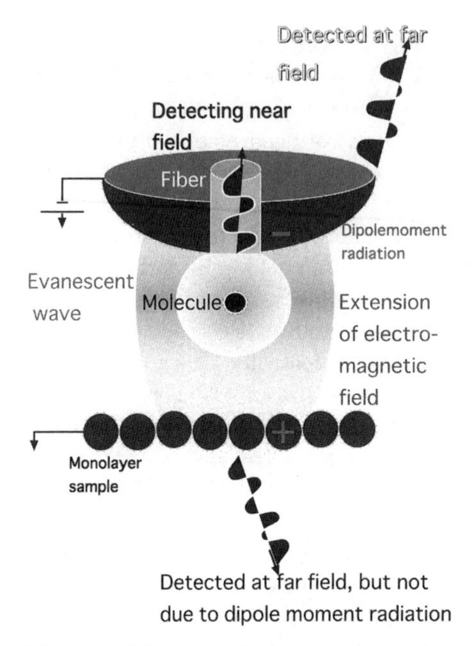

Fig. 1.8. Schematic diagram showing the evanescent field around a molecule. When a molecule is excited by a tunneling electron, it subsequently generates an evanescent field as it relaxes. This evanescent field exists only around the molecule whereas the electromagnetic field extends over the entire gap between the tip and the substrate. The diagram also compares evanescent field detection by a fiber tip and dipole moment radiation

regulated using force between the tip and sample. While the lateral extent of the electromagnetic field of photon emission from STM is also some tens of nanometers, as the extension of the plasmon is great along a metal surface, it is possible to considerably reduce the vertical size of the electromagnetic field by putting a molecule in the tunnel gap (Fig. 1.8) [135–142]. The molecule can be excited by a tunneling electron by holding a metal-coated fiber tip above it.

The effect of scattering a photon with a large $\boldsymbol{k}$ vector can be verified using two methods. In the first method, a molecule is placed on a metal substrate and a standard W STM tip is placed above the molecule (Fig. 1.9a). The molecule is excited by a tunneling electron and emits light when the molecule relaxes. A photon with a large $\boldsymbol{k}$ vector is scattered by the tip apex and turns into a photon with a smaller $\boldsymbol{k}$ vector that subsequently reaches a photon detector.

The second method uses an indium–tin-oxide (ITO) coated fiber tip [143] supported above a molecule which is excited by a tunneling electron. When this molecule emits light with a large $\boldsymbol{k}$ vector, in the ITO-coated fiber tip case, a photon with a large $\boldsymbol{k}$ vector is scattered and introduced into the fiber.

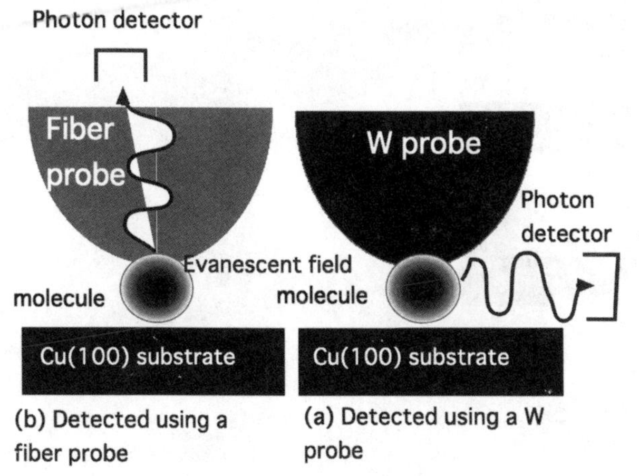

Fig. 1.9. A comparison of near-field detection using (**a**) a W tip, (**b**) a fiber tip. The evanescent field with a large $\boldsymbol{k}$ vector is scattered by the metal tip apex and evolves into a small $\boldsymbol{k}$ vector which subsequently reaches a photon detector (**a**). The evanescent field induces dipoles in the fiber tip and these dipoles propagate in the fiber tip to reach a photon detector (**b**)

In other words, a dipole is induced in the fiber and these dipoles propagate along the fiber. This dipole is also induced in the ITO layer, since ITO acts as an insulator for higher frequencies. A suitable experimental setup for studying an electromagnetic field at the nanometer scale is to study light emission from a molecule placed under an STM tip. In this system, the existence of a dipole moment is well defined. The dipole moment is not static but dynamic, and is responsible for the light emission. The direction of the dipole moment may then be manipulated parallel or perpendicular to the substrate surface. When the dipole moment is perpendicular to the surface, i.e., parallel to the tip–substrate direction, strong coupling with the plasmon results between the tip and the substrate surface and this plasmon turns into a radiative field. In contrast, when the dipole is parallel to the substrate surface, the coupling with the plasmon mode is weak so that the excited dipole is transferred to a neighboring surface molecule if there are many such molecules on the surface. This is a realistic situation when molecules are evaporated onto the surface.

1.6 Single Polariton Probe

As an example of a single polariton probe, consider the following experimental setup (Fig. 1.10) [144]. Two molecules are placed in nano-gaps between electrodes separated by a distance much smaller than an optical wavelength. Let us assume that these molecules are excited by a coherent electron wave with different phase. Then the phase of the electromagnetic field radiated from each molecule will be different so that the electromagnetic field cancels out at the far field.

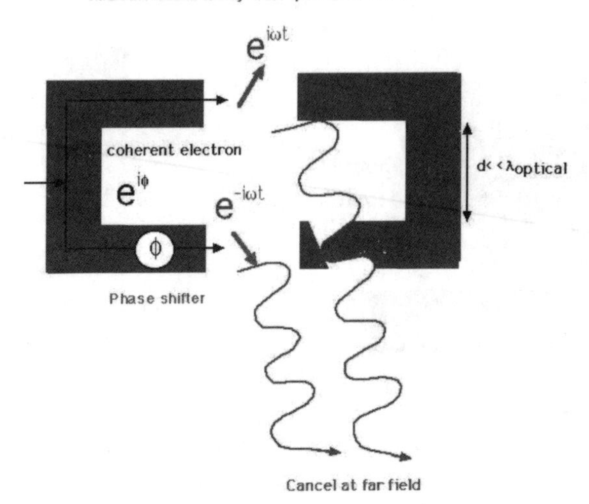

Fig. 1.10. Schematic view of prohibited photon emission. Two molecules are put in the gaps of the electric circuit and excited by a coherent electron. If the phase of the dipole moment of each molecule in the gap differs by π, the electromagnetic field cancels at the far field

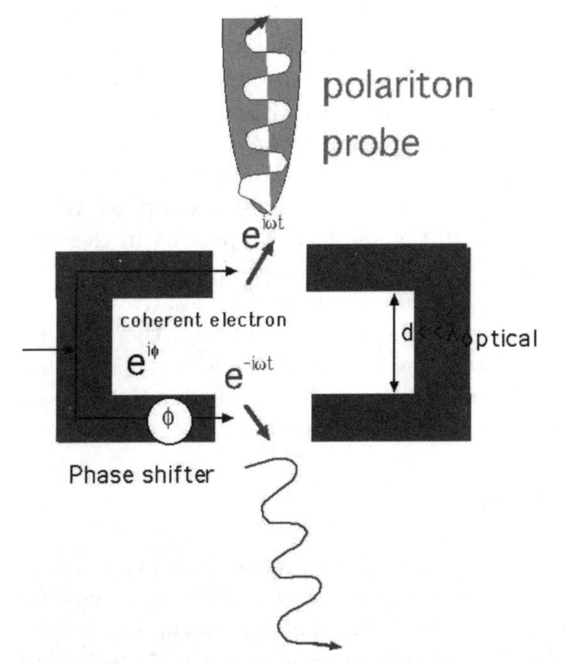

Fig. 1.11. Schematic view of allowed radiation. When a near-field probe is brought to one of the molecules so that it absorbs radiation from one side of the molecule, the electromagnetic field no longer cancels at the far field and the system emits light

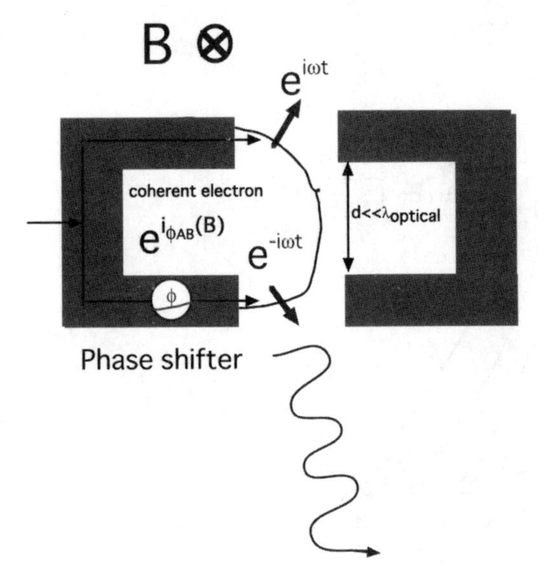

Fig. 1.12. Schematic view of a setup for observing photon emission. When the magnetic field is applied perpendicularly to this system, the phase of the electron is changed along a path shown in the drawing as a function of the magnetic field. The photon emission is thus modulated as a function of the magnetic field

Let us now imagine that we can place a probe above one of the molecules and thereby pick up a polariton (Fig. 1.11). As a consequence, the electromagnetic field radiated from the other molecule can reach the far field and this electromagnetic field can be observed.

This gedanken experiment can be extended. We consider a magnetic field applied vertically to the system (Fig. 1.12). By considering a loop along the electrode, the phase of the electron wave can be changed using the Aharanov–Bohm effect. The phase of two electromagnetic fields generated from each molecule can then be exchanged. The amplitude of the total electromagnetic field at the far field can thus be varied as a function of the magnetic field.

1.7 Coulomb Blockade and Connection to a Nonequilibrium Reservoir

In this section, we consider the meaning of the term 'observation', whereby the mode of an electromagnetic field as discussed above is connected to the macroscopic mode.

A Coulomb blockade should be understood in terms of the connection between a nanostructure and a macroscopic reservoir. Consider a double tunnel junction connected to a battery and an Ampermeter (Fig. 1.13). The system consists of two loops. The first loop includes the emitter, the central electrode

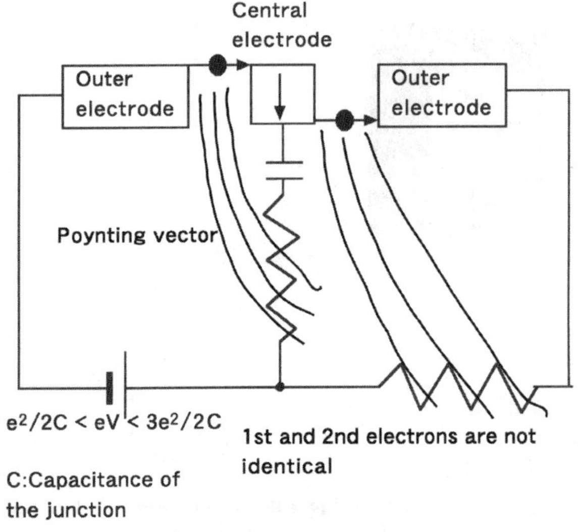

Fig. 1.13. Schematic view of a Coulomb blockade and electromagnetic field. When an electron tunnels through the first tunnel junction, the Poynting vector crosses the first resistor and this electron cannot be detected. When the first electron tunnels, another electron tunnels simultaneously through the second tunnel junction. Assuming that a photon is emitted during the first electron tunneling, the photon is detected by a photon detector

and a resistor connected to the central electrode via stray capacitance. The second loop includes the collector, the central electrode and another resistor connected to the central electrode. A Coulomb blockade arises when an initial electron is localized in the central electrode. A second electron cannot tunnel through the first potential barrier until the first electron leaves the central electrode. By observing the first electron, the second electron is blocked from tunneling. We consider the voltage range up to $e/2C$ where the charge of the central electrode is zero (Fig. 1.14). This suggests that the first electron resides virtually in the central electrode and cannot be observed from outside. This in turn leads to the consideration that the resistor in the first loop is connected to the reservoir but that this resistor is not connected to the detector. There is a correlation between the first and second electrons. When the first electron tunnels through the first potential barrier and reaches the central electrode, the Poynting vector crosses the resistor. This Poynting vector crossing is equated with observation of the electron. Put another way, when the first electron tunnels through the first potential barrier, the elec-

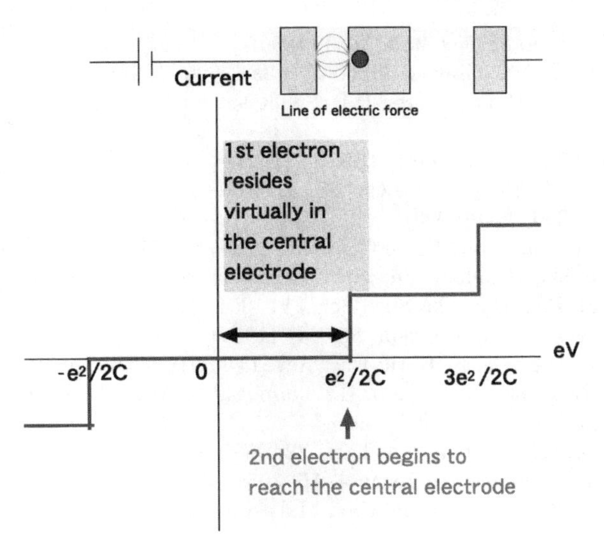

Fig. 1.14. Schematic view of a Coulomb staircase where the first electron resides virtually in the central electrode. The Poynting vector crosses a resistor that is connected to a reservoir but not to a detector so that we cannot observe the first electron in the Coulomb gap

tron which stayed in the second electrode should leave the central electrode simultaneously and this first electron passes through the resistor. According to this scheme, a Coulomb blockade is closely connected to the process of observation.

1.8 Structure of the Book

Chapter 2 describes photon emission experiments from single molecules using STM. The mechanism of single molecule photon emission is further described in Chap. 3. Fullerene molecules are commonly used for realizing nanoscale structures, and a detailed study of fullerene molecules is reported in Chap. 4. Chapter 5 describes the interaction between electrons and photons as demonstrated experimentally, whilst Chap. 6 provides the theoretical treatment of electron tunneling.

References

1. W.E. Moerner, M. Orrit: Science **283**, 1670 (1999)
2. B.C. Stipe, M.A. Rezaei, W. Ho: Science **280**, 1732 (1998)
3. J.K. Gimzewski: Photon emission from STM: Concepts' in: *Photons and Local Probes*, ed. by O. Marti and R. Möller (IBM, Netherlands 1995), pp. 189-208
4. R. Berndt: *Scanning Probe Microscopy*, ed. by R. Wiesendanger (Springer, Berlin 1998) p. 97

5. J.K. Gimzewski, C. Joachim: Science **283**, 1683 (1999)
6. T.A. Jung, R.R. Schlittler, J.K. Gimzewski: Nature **386**, 696 (1997)
7. J.K. Gimzewski, T.A. Jung, M.T. Cuberes, R.R. Schlittler: Surf. Sci. **386**, 101 (1997)
8. P. Sautet, C. Joachim: Chem. Phys. Lett. **185**, 23 (1991)
9. B.N.J. Persson, A. Baratoff: Phys. Rev. Lett. **68**, 3224 (1992)
10. A.P. Alivisatos: Science **271**, 933 (1996)
11. M. Welkowsky, R. Braunstein: Solid State Commun. **9**, 2139 (1971)
12. R. Berndt, J.K. Gimzewski, P. Johansson: Phys. Rev. Lett. **67**, 3796 (1991)
13. A. Adams, J. Moreland, P.K. Hansma: Surf. Sci. **111**, 351 (1981)
14. P. Avouris, B.N.J. Persson: J. Phys. Chem. **88**, 837 (1984)
15. K. Kuhnke, R. Becker, M. Epple, K. Kern: Phy. Rev. Lett. **79**, 3246 (1997)
16. N. Saito, F. Sato, K. Takazawa, J. Kusano, H. Okumura, T. Aida, T. Saiki, M. Ohtsu: Jpn. J. Appl. Phys. **36**, L896 (1997)
17. A.V. Zvyagin, J.D. White, M. Ohtsu: Opt. Lett. **22**, 955 (1997)
18. S. Mononobe, M. Ohtsu: J. Lightwave Technol. **15**, 1051 (1997)
19. M. Naya, R. Micheletto, S. Mononobe, R. Uma Maheswari, M. Ohtsu: Appl. Opt. **36**, 1681 (1997)
20. S. Mononobe, M. Naya, T. Saiki, M. Ohtsu: Appl. Opt. **36**, 1496 (1997)
21. S.-K. Eah, W. Jhe, T. Saiki, M. Ohtsu: Opt. Rev. **3**, 450 (1996)
22. R. Uma Maheswari, S. Mononobe, H. Tatsumi, Y. Katayama, M. Ohtsu: Opt. Rev. **3**, 463 (1996)
23. A. Zvyagin, M. Ohtsu: Opt. Commun. **133**, 328 (1996)
24. R. Uma Maheswari, S. Mononobe, M. Ohtsu: Appl. Opt. **35**, 6740 (1996)
25. R. Uma Maheswari, H. Kadono, M. Ohtsu: Opt. Commun. **131**, 133 (1996)
26. T. Tatsumoto, M. Ohtsu: J. Lightwave Technol. **14**, 2224 (1996)
27. S. Mononobe, M. Ohtsu: J. Lightwave Technol. **14**, 2231 (1996)
28. S. Mononobe, T. Saiki, T. Suzuki, S. Koshihara, M. Ohtsu: Opt. Commun. **146**, 45 (1998)
29. M. Ashino, M. Ohtsu: Appl. Phys. Lett. **72**, 1299 (1998)
30. T. Saiki, K. Nishi, M. Ohtsu: Jpn. J. Appl. Phys. **37**, 1638 (1998)
31. S. Mononobe, M. Ohtsu: IEEE Photonics Technol. Lett. **10**, 99 (1998)
32. A.V. Zvyagin, J.D. White, M. Kourogi, M. Kozuma, M. Ohtsu: Appl. Phys. Lett. **71**, 2541 (1997)
33. S. Mononobe, R. Uma Maheswari, M. Ohtsu: Optics Express **1**, 229 (1997)
34. T. Yatsui, M. Kourogi, M. Ohtsu: Appl. Phys. Lett. **71**, 1756 (1997)
35. T. Saiki, N. Saito, M. Ohtsu: Material Science and Engineering B **48**, 162 (1997)
36. M. Ohtsu: *Near-Field Nano/Atom Optics and Technology* (Springer, Tokyo, Berlin, New York 1998)
37. V.V. Polonski, Y. Yamamoto, J.D. White, M. Kourogi, M. Ohtsu: Jpn. J. Appl. Phys. **38**, L826 (1999)
38. M.B. Lee, M. Kourogi, T. Yatsui, K. Tsutsui, N. Atoda, M. Ohtsu: Appl. Opt. **38**, 3566 (1999)
39. K. Kobayashi, M. Ohtsu: J. Microscopy **194**, 249 (1999)
40. V.V. Polonski, Y. Yamamoto, M. Kourogi, H. Fukuda, M. Ohtsu: J. Microscopy **194**, 545 (1999)
41. H. Fukuda, Y. Kadota, M. Ohtsu: Jpn. J. Appl. Phys. **38**, L571 (1999)
42. K. Nikawa, T. Saiki, S. Inoue, M. Ohtsu: Appl. Phys. Lett. **74**, 1048 (1999)

43. T. Matsumoto, T. Ichimura, T. Yatsui, M. Kouroi, T. Saiki, M. Ohtsu: Optical Review **5**, 369 (1998)
44. T. Yatsui, M. Kourosgi, M. Ohtsu: Appl. Phys. Lett. **73**, 2090 (1998)
45. Y. Narita, T. Tadokoro, T. Ikeda, T. Saiki, S. Mononobe, M. Ohtsu: Appl. Spectroscopy **52**, 1141 (1998)
46. M. Ohtsu, H. Hori: *Near-Field Nano-Optics* (Kluwer Academic/Plenum Publishers, New York 1999)
47. R. Uma Maheswari, S. Mononobe, K. Yoshida, M. Yoshimoto, M. Ohtsu: Jpn. J. Appl. Phys. **38**, 6713 (1999)
48. M.B. Lee, N. Atoda, K. Tsutsui, M. Ohtsu: J. Vac. Sci. Technol. **B17**, 2462 (1999)
49. T. Matsumoto, M. Ohtsu, K. Matsuda, T. Saiki, H. Saito, K. Nishi: Apl. Phys. Lett. **75**, 3246 (1999)
50. Y. Yamamoto, M. Kourogi, M. Ohtsu, V. Polonski, G.H. Lee: Appl. Phys. Lett. **76**, 2173 (2000)
51. R. Wiesendanger, H.-J. Güntherodt: *Scanning Tunneling Microscopy II* (Springer-Verlag 1992)
52. D.V. Averin, K.K. Likharev: 'Single electronics: A correlated transfer of single electrons and Cooper pairs in systems of small tunnel junctions.' In: *Mesoscopic Phenomena in Solids*, ed. by B.L. Altshuler, P.A. Lee, R.A. Webb (North-Holland, Amsterdam 1991) pp. 173–272
53. T.A. Fulton, G.L. Dolan: Phys. Rev. Lett. **59**, 109 (1987)
54. P.J.M. van Bentum, R.T. Smorkers, H. van Kempen: Phys. Rev. Lett. **60**, 2543 (1988)
55. L.J. Geeligs, M. Peters, L.E.M. de Groot, A. Verbruggen, J.E. Mooij: Phys. Rev. Lett. **63**, 326 (1989)
56. M.H. Devoret, D. Esteve, H. Grabert, G.L. Ingold, H. Pothier, C. Urbina: Phys. Rev. Lett. **64**, 1565 (1990)
57. B. Su, V.J. Goldman, J.E. Cunningham: Science **255**, 313 (1992)
58. C. Schönenberger, H. van Houten, H.C. Donkersloot, A.M.T. van der Putten, L.J.G. Fokkink: Phys. Scr. T **45**, 289 (1992)
59. C.M. Fischer, M. Burghard, S. Roth, K. v. Klizing: Europhys. Lett. **28**, 129 (1994)
60. H. Nejo, M. Aono: Mod. Phys. Lett. B **6**, 187 (1992)
61. N. Shima: JRDC pre-research program, Abstract p. 17 (1994)
62. H. Nejo: Nature **353**, 640 (1991)
63. T.A. Jung, R.R. Schlittler, J.K. Gimzewski: Nature **386**, 696 (1997)
64. P.K. Tien, J.P. Gordon: Phys. Rev. **129**, 647 (1963)
65. N. Kroo, J.P. Thost, M. Völker, W. Krieger, H. Walther: Europhys. Lett. **15**, 289 (1991)
66. Y. Yamamoto, G. Björk, H. Heitmann, R. Horowicz: 'Controlled spontaneous emission in quantum well microcavities.' In: *Optics of Semiconductor Nanostructures* ed. by F. Henneberger, S. Schmitt-Rink, E.O. Göbel (Akademie Verlag, Berlin 1993) pp. 547–584
67. G. Björk, A. Karlsson, Y. Yamamoto: Phys. Rev. A **50**, 1675 (1994)
68. G. Björk: IEEE J. Quantum Electron QE **30**, 2314 (1994)
69. G. Björk, S. Pau, J.M. Jacobson, Y. Yamamoto: Phys. Rev. B **50**, 1736 (1994)
70. Y. Yamamoto, J.M. Jacobson, S. Pau, H. Cao, G. Björk: 'Exciton-polaritons in microcavities.' In: *Nanostructures and Quantum Effects* ed. by H. Sakai, H. Noge (Springer, Berlin, Heidelberg, New York 1994) pp. 157–164

71. J.M. Jacobson, H. Cao, G. Björk, S. Pau, Y. Yamamoto: 'Particle statistics and quantum dynamics in a semiconductor quantum-well microcavity.' In: *Extended Abstract of the 1994 International Conference on Solid State Devices and Materials*, (The Japan Society of Applied Physics, Japan 1994) pp. 133–135
72. H. Cao, J.M. Jacobson, G. Björk, S. Pau, Y. Yamamoto: Appl. Phys. Lett. **66**, 1107 (1995)
73. S. Pau, G. Björk, J.M. Jacobson, H. Cao, Y. Yamamoto: Phys. Rev. B **51**, 7090 (1995)
74. J.M. Jacobson, S. Pau, H. Cao, G. Björk, Y. Yamamoto: Phys. Rev. A **51**, 2542 (1995)
75. S. Pau, G. Björk, J.M. Jacobson, H. Cao, Y. Yamamoto: Phys. Rev. B **51**, 14437 (1995)
76. H. Cao, G. Klimovitch, G. Björk, Y. Yamamoto: Phys. Rev. Lett. **75**, 1146 (1995)
77. H. Cao, G. Klimovitch, G. Björk, Y. Yamamoto: Phys. Rev. B **52**, 12184 (1995)
78. S. Pau, G. Björk, J.M. Jacobson, Y. Yamamoto: Nuovo Cimento **17**, 1657 (1995)
79. G. Björk, S. Pau, J.M. Jacobson, H. Cao, Y. Yamamoto: Phys. Rev. B **52**, 17310 (1995)
80. G. Björk, Y. Yamamoto, H. Heitmann: 'Spontaneous emission control in semiconductor microcavities.' In: *Confined Electrons and Photons* ed. by E. Bernstein, C. Weisbuch (Plenum Press, New York 1995) pp. 467–501
81. G. Björk, Y. Yamamoto: 'Spontaneous emission in dielectric planar microcavities.' In: *Spontaneous Emission and Laser Oscillation in Microcavities* ed. by K. Ujihara, H. Yokoyama (CRC Press, Boca Raton, FL 1995) pp. 189–235
82. R.C. Liu, Y. Yamamoto: Phys. Rev. B **49**, 10520 (1994)
83. Y. Inoue, S. Kawata: Opt. Lett. **19**, 159 (1994)
84. T. Sugiura, T. Okada, Y. Inoue, O. Nakamura, S. Kawata: Opt. Lett. **22**, 1663 (1997)
85. Y. Inoue, S. Kawata: Opt. Commun. **134**, 31 (1997)
86. H. Furukawa, S. Kawata: Opt. Commun. **148**, 221 (1998)
87. T. Sugiura, S. Kawata, T. Okada: J. Microscopy **194**, 291 (1999)
88. H. Hatano, Y. Inoue, S. Kawata: Jpn. J. Appl. Phys. **37**, L1008, (1998)
89. N. Hayazawa, Y. Inoue, S. Kawata: J. Microscopy **194**, 472 (1999)
90. Y.Inoue: In: *Proceedings of the Twelfth International Conference on Fourier Transform Spectroscopy, Tokyo*, ed. by K. Itoh, M. Tasumi (Waseda University Press, Tokyo 1999)
91. R. Berndt, J.K. Gimzewski, P. Johansson: Phys. Rev. Lett. **67**, 3796 (1991)
92. R. Berndt, R.R. Schlittler, J.K. Gimzewski: J. Vac. Sci. Technol. **B9**, 573 (1991)
93. K. Ito, S. Ohyama, Y. Uehara, S. Ushioda: Surf. Sci. **324**, 282 (1995)
94. K. Ito, S. Ohyama, Y. Uehara, S. Ushioda: Appl. Phys. Lett. **67**, 2536 (1995)
95. S. Ushioda: Optoelectronics **10**, 193 (1995)
96. Y. Uehara, S. Ohyama, K. Ito, S. Ushioda: Jpn. J. Appl. Phys. **35**, L167 (1996)
97. K. Ito, S. Ohyama, Y. Uehara, S. Ushioda: Surf. Sci. **363**, 423 (1996)
98. Y. Uehara, K. Ito, S. Ushioda: Appl. Surf. Sci. **107**, 247 (1996)
99. S. Ushioda: Appl. Surf. Sci. **113/114**, 335 (1997)

100. M. Iwami, Y. Uehara, S. Ushioda: Rev. Sci. Instrum. **69**, 4010 (1998)
101. T. Tsuruoka, Y. Ohizumi, S. Ushioda, Y. Ohno, H. Ohno: Appl. Phys. Lett. **73**, 1544 (1998)
102. Y. Uehara, T. Fujita, S. Ushioda: Phys. Rev. Lett. **83**, 2445 (1999)
103. T. Tsuruoka, Y. Ohizumi, R. Tanimoto, S. Ushioda: Appl. Phys. Lett. **75**, 2289 (1999)
104. K.J. Ito, Y. Uehara, S. Ushioda, K. Ito: Rev. Sci. Instrum. **71**, 420 (2000)
105. Y. Uehara, A. Yagamai, K. Ito, S. Ushioda: Appl. Phys. Lett. **76**, 2487 (2000)
106. K. Kobayashi, M. Ohtsu: J. Microscopy **194**, 249 (1999)
107. J.K. Gimzewski, J.K. Sass, R.R. Schlittler, J. Schott: Europhys. Lett. **8**, 435 (1989)
108. R. Berndt, R. Gaisch, J.K. Gimzewski, B. Reihl, R.R. Schlittler, W.D. Schneider, M. Tschudy: Science **262**, 1425 (1993)
109. R. Berndt, J.K. Gimzewski: Surf. Sci. **269/270**, 556 (1992)
110. R. Berndt, J.K. Gimzewski: Phys. Rev. Lett. **67**, 3796 (1991)
111. R. Berndt, J.K. Gimzewski: Phys. Rev. **B48**, 4746 (1993)
112. J.K. Gimzewski, R. Berndt, R.R. Schlittler, A.W. McKinnon, M.E. Welland, T.M.H. Wong, P. Dumas, M. Gu, C. Syrkykh, F. Salvan, A. Hallimaoui: In *Near-Field Optics* ed. by D.W. Pohl, D. Courjon (Plenum, Amsterdam 1993) p. 333
113. R. Berndt, R.R. Schlittler, J.K. Gimzewski: J. Vac. Sci. Technol. **B9**, 573 (1991)
114. J.K. Gimzewski: In: *Photons and Local Probes* ed. by O. Marti, R. Moller (Plenum, Amsterdam 1995) p. 189
115. J.K. Gimzewski, B. Reihl, J.H. Coombs, R.R. Schlittler: Z. Phys. B Condensed Matter **72**, 497 (1988)
116. J.H. Coombs, J.K. Gimzewski, B. Reihl, J.K. Sass, R.R. Schlittler: J. Microscopy **152**, 325 (1988)
117. P. Johansson, R. Monreal, P. Apell: Phys. Rev. **B42**, 9210 (1990)
118. R.W. Rendell, D.J. Scalapino: Phys. Rev. **B24**, 327 (1981)
119. B. Laks, D.L. Mills: Phys. Rev. **B20** 4962 (1979)
120. B. Laks, D.L. Mills: Phys. Rev. **B21** 5175 (1980)
121. C.A. Neugebauer, M.B. Webb: J. Appl. Phys. **33**, 74 (1962)
122. J. Lamber, S.L. McCarthy: Phys. Rev. Lett. **37**, 923 (1976)
123. D. Hone, B. Mühlschlegel, D.J. Scalapino: Appl. Phys. Lett. **33**, 203 (1978)
124. R.W. Rendell, D.J. Scalapino, B. Mühlschlegel: Phys. Rev. Lett. **41**, 1746 (1978)
125. J.R. Kirtley, T.N. Theis, J.C. Tsang, D.J. MiMaria: Phys. Rev. **B27**, 4601 (1983)
126. D.L. Abraham, A. Veider, C. Schöenberger, H.P. Meier, D.J. Arent, S.F. Alvarado: Appl. Phys. Lett. **56**, 1564 (1990)
127. E. Flaxer, O. Sneh, O. Cheshnovsky: Science **262**, 2012 (1993)
128. D.H. Waldeck, A.P. Alivisatos, C.B. Harris: Surf. Sci. **158**, 103 (1985)
129. M. Nirmal, B.O. Dabbousi, M.G. Bawendi, J.J. Macklin, J.K. Trautman, T.D. Harris, L.E. Brus: Nature **383**, 802 (1996)
130. A. Otto, I. Mrozek, H. Grabhorn, W. Akemann: J. Phys. Condens. Matter **4**, 1143 (1992)
131. H. Nejo, Z.-C.Dong: 'Interactions of electrons and electromagnetic fields in a single molecule.' In: *Optical and Electronic Processes of Nano-Matter* ed. M. Ohtsu (KTK Scientific Publishers, Tokyo 2001) pp. 123–146

132. M. Tsukada, T. Shimizu, K. Kobayashi: Ultramicroscopy **42/44**, 360 (1992)
133. R.I. Hall, F.M. Read: 'Molecular spectroscopy by electron scattering.' In: *Electron–Molecule Collisions* ed. by I. Shimamura, K. Takayanagi (Plenum, New York 1984)
134. E.W. McDoniel: *Atomic Collisions: Electron–Photon Projectiles* (ISBN 0-471-85307-0) (Wiley, New York 1989)
135. I.B. Berlam: *Handbook of Fluorescence Spectra of Aromatic Molecules* (Academic Press, New York 1965)
136. M. Gouterman: In: *The Porphyrins*, ed. by D. Dolphin (Academic Press, New York 1978) Part A, pp. 1–165
137. R.M. Dickson, A.B. Cubitt, R.Y. Tsien, W.E. Moerner: Nature **388**, 355 (1997)
138. Z. Wu, T. Nakayama, S. Qiao, M. Aono: Appl. Phys. Lett. **73**, 2269 (1998)
139. D. Eastwood, M. Gouterman: J. Mol. Spectrosc. **30**, 437 (1969)
140. D. Dolphin: *The Porphyrins*, Vol. 3 (Physical Chemistry, Part A) (Academic Press, New York 1978)
141. T. Kobayashi, D. Huppert, K.D. Stranb, P.M. Rentzepis: J. Chem. Phys. **70**, 1720 (1979)
142. W.E. Moerner, M. Orrit: Science **283**, 1970 (1999)
143. S. Sasaki, T. Murashita: Jpn. J. Appl. Phys. **38**, L4 (1999)
144. H. Hori: 'Electronic and electromagnetic properties in nanometer scales.' In: *Optical and Electronic Processes of Nano-Matter* ed. by M. Ohtsu (KTK Scientific Publishers, Tokyo 2001) pp. 1–55

2 STM-Induced Photon Emission from Single Molecules

Z.-C. Dong, T. Ohgi, D. Fujita, H. Nejo, S. Yokoyama, T. Terui, S. Mashiko, and T. Okamoto

The investigation of single quantum structures lies at the frontier of modern chemistry, physics, and nanoelectronics. A single molecule is a simple quantum system for studying electron–electron and electron–photon interactions and other quantum mechanical effects. The quantum nature of molecular systems makes single-molecule studies important not only for a fundamental understanding of basic phenomena underlying bulk molecular sample behavior, such as in organic electronics, but also for shaping emerging concepts and principles of molecular nanotechnology. For instance, over the last decade, organic thin film science has been developing at an unprecedented pace with attractive technological applications in organic light-emitting diodes, dye lasers, organic transistors, liquid-crystal displays, and photovoltaic cells. However, even with these technological breakthroughs, there remain deep questions about the underlying physics of charge and exciton transport, energy absorption–transfer–decay processes, optical interactions, and the nature of electronic states within these bulk molecular materials. Molecular-scale investigations will clarify the fundamental mechanisms of these phenomena and processes and thus offer guidelines for designing better molecular materials and devices.

Single molecular studies are also driven by the emerging field of molecular scale electronics, which aims to develop electronic components such as wires, switches, amplifiers, and transistors via the use of just a few molecules or even a single supramolecular assembly, and to integrate these into a device. The field involves many challenging unsolved issues such as the electrical addressing of molecules and circuit design (e.g., input/output interconnects and clock frequency). We are still at the materials stage, seeking to understand at a fundamental level what is going on in molecules by measuring their transport and optoelectronic characteristics. To date, most transport phenomena measured seem to be of contact-type rather than a true conductivity of the molecule itself, except for those related to internal motion (e.g., vibronic and rotational transitions). Scanning tunneling microscopy provides controlled two-terminal measurements, and its development together with other techniques should offer new experimental approaches for probing electron–photon interactions in, and electron transport through individual molecules. Eventually, this may lead to the determination of true molecular conductance, a measure of the electronic transparency of a single molecule.

Single-molecule studies comprise three aspects: microscopy in the space domain, spectroscopy in the frequency domain, and fluorescence decay in the time domain. To understand single-molecule behavior in a nano-environment, it is necessary to describe the electromagnetic radiation in the nanoscopic system. Scanning probe techniques (e.g., scanning tunneling microscopy) allow study of individual molecules on surfaces, while optical techniques (e.g., ultrasensitive fluorescence spectroscopy) enable characterization of single-molecule behavior in complex condensed environments. Single-molecule measurements make it possible to unravel the normal ensemble averaging that occurs when a large number of molecules are probed simultaneously. This allows direct observation of the true distribution and time-resolved information about a given property (structure and functions) of a molecule, or many single molecules but on a one-by-one basis [1].

The invention of the scanning tunneling microscope (STM) has revolutionized our way of dealing with individual atoms and molecules. It can provide atomically resolved information on the conformation and adsorption nature of individual molecules on a surface. Yet it can do more than just observe the nanoworld. It can also be used to manipulate atoms and molecules and to construct supramolecular assemblies and functional molecular machines. Furthermore, combined STM and optical techniques are particularly attractive for single-molecule studies. It becomes possible not only to image individual molecules on a surface, but also, through probing photons emitted via inelastic processes, to gain insights into the tunneling regime, e.g., specific characteristics of diverse molecule and substrate excitations as well as modified optical properties due to quantum confinement [2–4].

Inelastic processes excited by tunneling electrons carry chemically specific information on molecular systems, but are very difficult to detect by conventional scanning tunneling spectroscopy alone due to their small contribution to the total tunneling current [5]. Tunneling of electrons or holes from an STM tip provides an extremely localized source of low-energy carriers (only a few eV), which offers resolution down to the subnanometer scale for photon emission mapping. Spatial mapping of optical signals may enable us to add 'color' to STM images and perform chemical mapping on surfaces. Such high resolution spectral photon maps will give access to the electromagnetic coupling between the tip and sample and its variation even on the atomic scale. Another advantage of STM is its freedom to vary experimental parameters on well-characterized surfaces, thus permitting detailed investigation of the processes involved.

The atomic resolution of a scanning tunneling microscope arises from the extremely high sensitivity of tunneling currents to the small tail overlap between the wavefunctions of both tip and substrate across the tunneling barrier. How does the molecule couple with the electrodes and what happens optically when a single molecule is inserted into this subnanometer gap and the molecular junction is biased (Fig. 2.1)? This is the issue we shall address

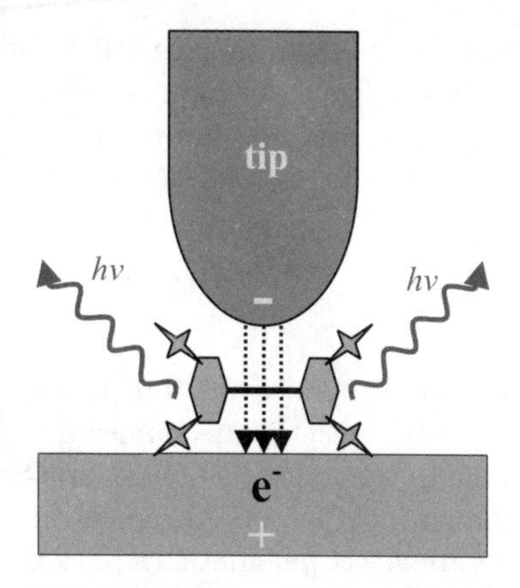

Fig. 2.1. Schematic view of a molecule sandwiched between tip and substrate. Light is emitted due to excitation by tunneling electrons

in the present chapter. Following a brief review of STM-induced photon emission from metal and semiconductor surfaces, we shall report our research on tunneling-electron-induced light emission from porphyrin molecules. General concepts and principles involved in STM-induced photon emission phenomena, experimental setup, factors affecting photon emission, light emission mechanisms, and current challenges in the field will be discussed.

2.1 Past and Present

Before we proceed to the discussion of STM-induced photon emission from single molecules, we shall briefly review the field of light emission induced by tunneling electrons on semiconductor and metal surfaces [2,4,6].

2.1.1 Semiconductor Surfaces

Photon emission stimulated by tunneling electrons can be traced back to Lambe and McCarthy in 1976 on metal–oxide–metal solid-state junctions [7]. The first experimental evidence of STM-induced light emission was reported by Gimzewski et al. from Ta and Si(111) surfaces [8] and soon afterwards, enhanced light emission from Ag films was also observed [9]. These results showed similarities to conventional inverse photoemission [10] with additional spectral features assigned to optical transitions from field-induced states. STM is also used for the investigation of local cathodoluminescence (CL) in GaAs/$Ga_{1-x}Al_xAs$ heterostructures [11–13] and CdS surfaces [14] in which emission arises from optical transitions between band edges. The polarization effects of STM-induced photon emission have also been reported for GaAs [15] and Si(111) [16]. There are also reports using a conductive and transparent indium–tin-oxide (ITO) tip for both carrier injection and photon collection [17], the so-called near-field detection (to be contrasted with far-field detection using a lens system located far away from the light source). Recent progress in the field has provided photon maps with spatial resolution down to the atomic scale for Si(111) [18] and hydrogen-passivated Si(100) [19] surfaces.

2.1.2 Metal Surfaces

On the other hand, research on STM-induced photon emission from metal surfaces has also received a great deal of attention because the mechanism involved in the emission process is new and specific to the nanometer proximity and electromagnetic coupling of the tip to the surface [2,4,6]. An energetic electron impinging on a metal surface can generate electromagnetic radiation via dipolar resonance mechanisms such as transition radiation and surface plasmon excitation. An STM tip can be used as an electron source for generating fluorescence on metal surfaces in both the far-field emission regime and regimes of proximity field emission and tunneling. At high bias voltages ($\geq 100\,\mathrm{V}$), emission phenomena were found to be similar to those due to conventional electron guns [20]. However, in the proximity field-emission regime ($\leq 40\,\mathrm{V}$) and in the tunneling regime ($\leq 5\,\mathrm{V}$), a new and interesting phenomenon was observed [21]. Instead of the absence of photon emission by conventional electron guns for such low electron energies, intense photon emission occurs with redshifted optical spectra and an enhanced quantum efficiency up to $\sim 10^{-4}$ photons/electron for a W tip on Ag. A universal intensity maximum for bias voltages between 3–4 V is observed in the tunneling regime on all metallic surfaces. This has been attributed to competition between the initial increase in allowed channels for photon emisison and the subsequent field-enhancement weakening due to the vertical tip retraction in that region [22]. In an STM configuration, the tip–sample distance is in the nanometer region, much shorter than the wavelength of the emitted light. This results in localized electromagnetic coupling between tip and sample and hence new emission features.

Different single-crystal metal surfaces have been shown to exhibit different emission characteristics [20], but the spatial resolution of chemical signals has not been fully investigated. The localization of photon emission to a small area beneath the tip promises a form of chemical microscopy, because the photon signal is related to the local chemical environment, and chemical information can be extracted from the complicated photon emission signals [18,23]. Strictly speaking, the spectroscopy of emitted light is the only way to identify unknown metallic structures. Nevertheless, since different metals have different optical properties, their emission efficiencies can vary over two orders of magnitude and photon mapping can thus be effectively used to distinguish known regions on flat surfaces. Analysis of photon maps is based on simultaneous topographic, bias-dependent imaging in combination with spectroscopic data. A contrast mechanism thus formulated can help to reveal various factors that determine the local efficiency of photon emission and to gain further insight into the physical origin of factors governing inelastic tunneling excitation of photons in mesoscopic systems.

2.1.3 Mechanisms

Theoretically, photon emission from metal surfaces was attributed to radiative decay of plasmons for solid state tunneling junctions and STM-induced photon emission. These plasmons are excited either by inelastic tunneling (IET) [24,25] or by hot-electron thermalization (HET) [26].

For IET excitation, a tunneling electron gives away a fraction of its energy to a plasmon that is subsequently converted to a photon, as shown in Fig. 2.2a and b. Tunneling electrons directly excite the charge density oscillation between the tip and surface, i.e., the local (interface) plasmon modes. The local plasmons transfer their energy to photons that are then emitted.

In HET, an electron tunnels through a barrier elastically and then loses its energy to the sample by thermalization (or hot-electron decay) via plasmon generation. The inelastic event, which leads to photon emission, occurs during tunneling for IET but after tunneling within the sample for HET. Hot-electron injection is still debated as one of the mechanisms for photon emission from metal–oxide–metal junctions, whereas in STM configurations, inelastic tunneling excitation is believed to be dominant in the photon emission process from metal and indirect semiconductor surfaces. Hot-electron excitation is negligible compared with IET excitation in terms of high intensities and lack of significant bias dependence (suggestive of a resonant process

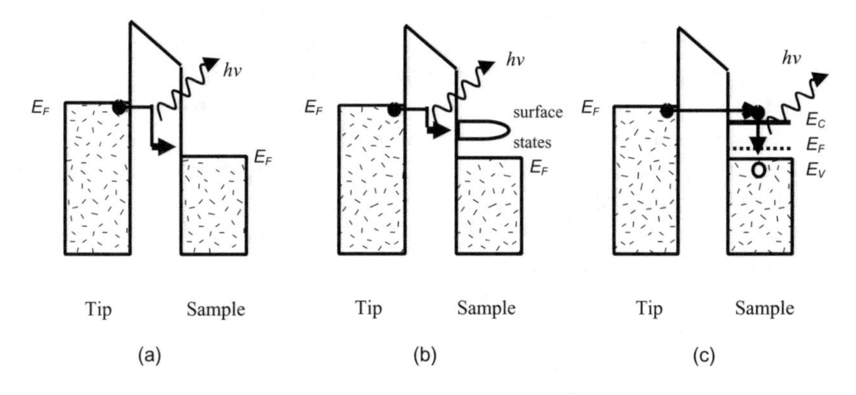

Fig. 2.2. Schematics of mechanisms for STM-induced photon emission: (**a**) excitation of localized surface plasmons (LSP) on metal surfaces via inelastic tunneling (IET), the so-called tip-induced plasmon (TIP) mode, (**b**) radiative decay via inelastic tunneling into surface states on metal and semiconductor surfaces, (**c**) conventional luminescence via electron–hole pair recombination through elastic tunneling on direct band gap semiconductor surfaces

during tunneling) as well as consistent fluorescence spectra between theory and experiment.

Barrier height measurements show almost the same value for both elastic and inelastic tunneling [27]. This is somewhat unexpected because one would expect the inelastic barrier height to be larger than the elastic one by $\sim h\nu/2$. An electron that tunnels inelastically is therefore thought to traverse most of the barrier elastically and undergo an inelastic event only toward the end of tunneling. On the other hand, for STM-induced light emission from semiconductor surfaces, in particular direct band gap materials such as *p*-type GaAs [11], hot-electron decay within the semiconductor is the primary mechanism for luminescence. Photon emission is due to elastic tunneling of electrons from the tip into the GaAs conduction band and subsequent recombination with holes in the valence band, i.e., electron–hole pair recombination, as shown in Fig. 2.2c.

There is increasing interest in the investigation of light emission from surface states associated with either intrinsic surfaces or adsorbed atoms or molecules. This is due on the one hand to an ongoing effort to understand the electronic properties of a surface with or without adsorbates on a very local scale, and on the other hand, the lack of a well-established physical picture for the light emission process. The mechanism shown in Fig. 2.2b, though favored, is not the only one available for interpreting light emission [19]. An HET mechanism via surface states is also possible.

Two models have been proposed to explain STM-induced photon emission from metal surfaces [4]. The model due to Persson and Baratoff [28] assumes that tunneling occurs from an *s*-like orbital at the tip apex to a spherical free-electron-like metallic particle. This model predicts that enhanced photon emission occurs if tunneling electrons excite the dipole plasmon resonance of the metal particle, while direct transition radiation is found to be low. Furthermore, the dipolar plasmon excitation is predominantly caused by the inelastic tunneling process (with a quantum efficiency up to 10^{-3}) in comparison with hot-electron decay (10^{-5}). Intensive photon emission has been interpreted as arising from the radiative decay of dipolar plasmons created by the close proximity of a tip to a surface.

Another model (a fashionable electrostatic one) due to Johansson et al. [29] takes into account the role of a tip, viewing tunneling as occurring between a metallic nanosphere and a planar metal surface. The classical electromagnetic response of the system is calculated to obtain the strength of field fluctuations that are responsible for photon emission. The emission intensity is determined from the product of the power spectrum of the tunneling current and that of the enhanced electric field induced by the tip. Light emission is resonantly enhanced due to the formation of local coupled plasmons, the so-called tip-induced plasmon (TIP) modes in the cavity formed between the tip and sample. All modes of the electromagnetic field inside the tunneling gap are strongly coupled to the collective motion of the electron gas inside the tip and sample, and hence, all of them should be called plasmons. These plasmons decay into photons outside the gap, in other words, the decoupling occurs far away from the metal surface in the free space where the modes of the free electromagnetic field can be called photons. The resonance occurs at the frequency at which half a wavelength of the TIP modes fits into the cavity between the tip and sample. Compared with light emission in vacuum, the enhancement could be as large as 4–5 orders of magnitude. The quantum efficiency is typically 10^{-4} photons/electron and is as high as 10^{-3} for an Ag tip on an Ag surface [14].

These TIP modes are similar to the collective charge density oscillations observed at two metal–dielectric interfaces in close proximity [30]. The lowest-order dipolar mode gives rise to a high probability of inelastic tunneling and photon emission. For an enhanced electromagnetic field under the tip, tunneling electrons that have a wave vector parallel to the dipole mode can couple efficiently via the IET mechanism. A recent study on Ag/Si(111) [31] suggested that higher-order multipole modes are responsible for higher-energy spectral peaks. If sample and tip are made of the same material, the resonance frequency of localized plasmons between tip and sample is given approximately by [25,31]

$$\omega_n = \omega_{\mathrm{p}} \left(\frac{q_n d}{2} \right)^{1/2} = n^{1/2} \omega_{\mathrm{p}} \left(\frac{d}{8R} \right)^{1/4} , \tag{2.1}$$

where n is the order of the multipole mode (roughly speaking, $n = 1, 2, 3$ correspond to the dipole, quadrupole, and hexapole mode, respectively), ω_{p} the bulk plasma frequency of the free electron gas, q_n the wave number of the resonance multipole mode [$q_n = nq_1 \sim n(2\mathrm{Rd})^{-1/2}$], d the tip–sample distance, and R the tip radius. Using the IET–TIP mechanism, Berndt et al. have further demonstrated that photon spectra can be modeled fairly well for Cu, Au, and (to a lesser extent) Ag surfaces regarding the positions of maxima, cutoff wavelength, and signal intensities [32]. The intensity of plasmon excitation was also evaluated by Tsukada et al. [33] using the first-principles calculation within the framework of surface plasmon excitation via inelastic tunneling.

2.1.4 Factors Affecting STM-induced Photon Emission and Modes of Measurements

STM-induced photon emission is determined by a variety of factors. In addition to the tunneling current and bias voltage, the IET–TIP model implies that the tip–sample distance, tip radius, and work functions and dielectric properties of the tip and sample all affect the local TIP modes. In order to separate the factors that affect photon emission, STM-induced light emission experiments are mainly carried out in three different modes: photon mapping, optical emission spectroscopy, and isochromat spectroscopy [4]. The lateral extent of these IET–TIP modes on flat surfaces is theoretically determined to be below $\sim$ 10 nm [34] (the localization length of localized plasmons is equal to $(Rd)^{1/2}$ [35]), and hence we can expect contrast in photon maps for surface features with geometric dimensions on that scale. On the other hand, electronic properties determine the initial and final density of states available for inelastic processes, and they will modify the branching ratio between elastic and inelastic channels and hence photon emission. Consequently, one expects local variations in such factors to give rise to contrast in maps of the integral photon intensity (i.e., photon maps). Analysis of photon maps has to take into account both geometric and electronic aspects via bias-dependent topography and spectroscopy. Photon maps [36] and emission spectra [42] with atomic spatial resolution were recently reported for the Au(110) surface with different features for over-row and between-row atoms. Polarization effects of photon emission have also been investigated for Co films and were found to be related to magneto-optical effects [38,39].

2.1.5 The Debate on Mechanisms

Although the theory of the tip-induced plasmon mode via inelastic tunneling (IET–TIP) is generally accepted for the modeling of STM-induced photon emission on metal surfaces, there is still some debate about the retardation effect and even the role of localized surface plasmons. A recent fully-retarded calculation by Johansson [40] indicates that the inclusion of retardation effects does not change the basic picture of STM-induced photon emission described above but does change spectral positions at a quantitative level. In general, retardation effects become important when the sample or tip is a good conductor (e.g., Ag). The energy dissipation is then rather small in the near-field zone and a localized mode will instead be damped because some energy propagates away from the near-field zone. Retardation therefore leads to an additional redshift of the resonance frequency. For Au and Cu samples, the retardation effect is negligible at the resonance frequency ($\sim$ 2.1 eV for Au and 2.0 eV for Cu) and the emission intensity is essentially unchanged.

Other theories have also been proposed to explain STM-induced photon emission. Since the tunneling distance, the localization length of localized plasmons, and the characteristic plasmon length of the electron gas in the tip and sample all lie in the nanometer (or subnanometer) range and are quite comparable to each other, a precise theoretical model of inelastic electron tunneling should take into account the spatial dispersion of the dielectric function of the electron gas, i.e., the non-local effects. Ushioda and coworkers [41,37] developed a theory for visible light emission from STM on the basis of the theory designed for solid-state tunnel junctions. Their theory concerns the differences between the branching ratios of surface plasmon polaritons that propagate along the surface (in other words, are delocalized) and those that are localized under the tip (TIP modes) and emit light directly. A radiation mechanism through the Cherenkov effect (or 'inverse Landau damping') was also proposed to explain the resonant excitation of interface plasmons by tunneling electrons via phase matching between the plasmon phase velocity and the Fermi velocity of electrons [43].

Recently, Xiao opened a debate about the importance of surface plasmons in the process of photon emission [44–47]. Specifically, does light emission require the excitation and de-excitation of surface plasmons? Using the point dipole model, he proposed the following simplified Hamiltonian to explain light emission from the tip–sample junction:

$$H = H_0 + eV \cos \omega t \,, \tag{2.2}$$

where H_0 is the Hamiltonian of the electron moving through the potential barrier between tip and sample, V is the bias voltage applied, and ω is the characteristic frequency of the tunneling current. According to Xiao, a system with such a simplified Hamiltonian is able to demonstrate that the coupling modes inside the tip–sample cavity can be excited by the fluctuation of tunneling currents in analogy with the situation in cavity quantum electrodynamics [48]. These excited coupling modes are responsible for light emission in STM; the use of local surface plasmons is not needed to explain the observed photon emission. It is worth noting that the interaction term in (2.2) is constant in space and is not very suitable for inducing transitions due to the fluctuation of electric fields [45]. Nevertheless, these arguments indicate that the theory of photon emission induced by tunneling electrons is not yet well established.

2.1.6 Molecules on Surfaces

In contrast with the extensive investigations of STM-induced light emission on semiconductor and metal surfaces, only a few scattered reports have been found on photon emission from molecules adsorbed on metal surfaces. The scarcity of data on molecules is mainly due to the difficulty in obtaining reliable photon maps and optical spectra without modifying or damaging the molecules.

The first (and probably by far the best defined) photon map with molecular resolution was obtained by Berndt et al. [49] with a simultaneous topograph ($V_s = 2.8\,\text{V}$, $I_t = 4.4\,\text{nA}$) for a monolayer of C_{60} on Au(110). The C_{60} molecules appear as a hexagonal array of individual photon emitters with diameter $\sim 4\,\text{Å}$ and separation $\sim 10\,\text{Å}$. The high symmetry and stability of C_{60} molecules are the key to the success of this experiment. The authors observed more intense emission when the STM tip was positioned above a C_{60} molecule and went on to claim that the molecule itself was the primary photon source. However, due to the lack of optical spectra, it is still not conclusively demonstrated that the observed molecular contrast on photon maps involves molecular fluorescence from C_{60}. The C_{60} molecular layers could act like some kind of grating which modifies the spatial distribution of light emission from the Au(110) surface. In any case, it is clear that the molecules couple strongly with the electromagnetic modes of the cavity between the tip and sample and thus play an active role in the emission process.

Excited molecules are known to couple with the plasmons of a metal surface [50]. When a molecule is sandwiched between tip and substrate, molecular fluorescence may be enhanced via a similar mechanism to that in surface-enhanced Raman spectroscopy (SERS) [51]. Electromagnetic confinement may mediate and amplify emission from the coupled tip–molecule–substrate system. To see how emission is modified by such confinement, it would be desirable to obtain the optical spectra for such a configuration and compare them with photoluminescence and electroluminescence studies.

Berndt et al. also reported molecularly resolved photon maps of individual anthracene molecules on Ag(110) [4] ($V_s = -2.6$ V, $I_t = 30$ nA), but there was no simultaneously acquired topography image to compare with and the characteristics of molecules after photon emission remain unidentified. According to Berndt, the molecular emission is more intense than that from the Ag substrate. The emission intensity from anthracene molecules is strongly dependent on bias polarity whereas emission from the Ag substrate is relatively insensitive to it.

The nonradiative damping of molecular excited states by metal surfaces is thought to drastically quench fluorescence when molecules are in close proximity to metallic surfaces [52]. To suppress the quenching of molecular emission on metal surfaces, Flaxer et al. [53] used the transparent and conducting thin film of indium–tin-oxides (ITO) as the substrate for the deposition of submonolayers of dichloroanthracene or coumarin molecules. No STM images, photon maps, and optical spectra were reported there, so it remained unclear what the surface looked like and what happened to the molecules under a tunneling current up to 50 nA and a bias voltage above 2.6 V. Nevertheless, the bias and coverage dependence of photon yields from the squeezable tunnel junction does strongly suggest the presence of light emission associated with molecular fluorescence excited by inelastic electron tunneling, although the 10^{-5}% quantum efficiency estimated is still very low. The poor emission efficiency from molecules on metal and semiconductor surfaces is due to the fast deexcitation processes governed by energy transfer to nonradiative surface excitations (Auger processes) [54].

Another STM-induced light emission from molecules in air was carried out for monolayers of copper phthalocyanine (CuPc) molecules on Au(111) [43]. Pleated structures of CuPc molecules were resolved on both STM topographs and photon maps but with a poor resolution, partly due to the large current used (10 nA). On the basis of photon maps and bias dependence of emission intensities, the authors viewed each CuPc molecule as a source of secondary plasmons upon interaction with the interface plasmons excited by tunneling electrons. The periodicity of the molecular structure should be preserved in the photon map, but the positions of emission maxima do not necessarily correspond to the positions of individual molecules. It is worth noting that measurements in air may suffer from complications caused by contamination and the resultant altered tunneling behavior as well as electrical breakdown and surface modifications at high bias voltages [4,55].

Recently, STM-induced photon emission has been reported for submonolayers of Cu-TBP porphyrin molecules on Cu(100) in ultrahigh vacuum via near-field detection from a conductive fiber tip [56]. However, it is difficult there to discern how the molecules are related to the correlation between the photon map and simultaneous topograph acquired under tunneling conditions of 6.0 V and 5.0 nA, since the molecules are likely to be damaged in such a high-current, field-emission regime. The authors estimated the quantum efficiency to be about 2×10^{-5} photons/electron and claimed that Cu-TBP porphyrin molecules contribute to light emission on the basis of the enhancement with respect to the noble metal surface and a faint peak at 2.2 eV in optical spectra. Light emission with a higher quantum efficiency appears quite possible if a designed molecular chromophore such as hexa-butyl decacyclene (HB-DC) is used [57].

As discussed above, the theory of photon emission induced by tunneling electrons on metal surfaces is still not yet well established. The situation becomes even more complicated and more difficult to study when molecules are sandwiched between the tip and surface. The acquisition, distinction, and separation of molecular fluorescence with respect to photon emission from the tunnel junction or metal surfaces is not the only experimental challenge. In addition, the mechanism whereby electrical energy is transformed into light energy via molecules remains a puzzling issue. For tip–molecule–substrate structures, the details of this electroluminescent process – energy absorption or excitation, energy transfer, and emission – are still far from clarified. The focus is on the coupling of molecular electronic states with the local electromagnetic field, in particular with regard to the precise way in which a molecule couples with tunneling electrons, gets excited, and gives out light.

We have recently initiated an investigation of organic molecules such as porphyrins bound onto a metal surface. This work is conducted with the following objectives:

- to understand the mechanism and underlying physics involved in the emission process, in particular the electron transport through a single molecule and electron–photon conversion,
- to explore the ultimate spatial and spectral resolution of photon emission for chemical mapping and molecular recognition on surfaces,
- to provide guidelines for improving the quantum efficiency of electroluminescence (EL) via molecules for future molecular electronics.

We report here our preliminary results on STM-induced photon emission from Cu(100) covered by porphyrin molecules, and present a wavelength-resolved optical spectrum that appears consistent with molecular vibrational transitions. Some general issues on experiments, molecules, and tunneling are also discussed throughout the paper.

2.2 STM-Induced Light Emission from Porphyrin Molecules

2.2.1 Experimental Setup

It is no easy task to perform a successful STM-induced light emission experiment. It requires a well-defined sample, good tip, efficient photon collection system, and highly sensitive photon detection system. Photomultipliers (PMT), grating spectrometers, intensified diode array detectors, or intensified CCDs are usually used to detect and spectrally resolve the photon emission. Since the photon intensity is very low due to the low quantum efficiency from the STM tunnel junction, a large collected solid angle becomes one of the most desirable features in designing a photon collection system. These systems can be divided into three main categories: mirror, lens, and fiber setups. The ellipsoidal or parabolic mirror system focuses light from the tip region onto the focal point outside the UHV close to a viewport [31,58]. This setup covers a large solid angle up to 61% of 2π, but is relatively difficult to make and align. The fiber bunch system is relatively easy to make and can also offer a large collection efficiency but it is difficult to make alignments [19]. The optical fiber tip system can be classified in this category with its unique feature of collecting photons at the near field [17]. Lens systems are convenient to use and are the most widely used setup nowadays in STM-induced photon emission studies [9,18,42]. However, they suffer from a small solid angle of collection. This is also the setup we use for photon collection (Fig. 2.3), the so-called far-field detection, since the distance between lens and light source is much greater than the wavelength of light emitted.

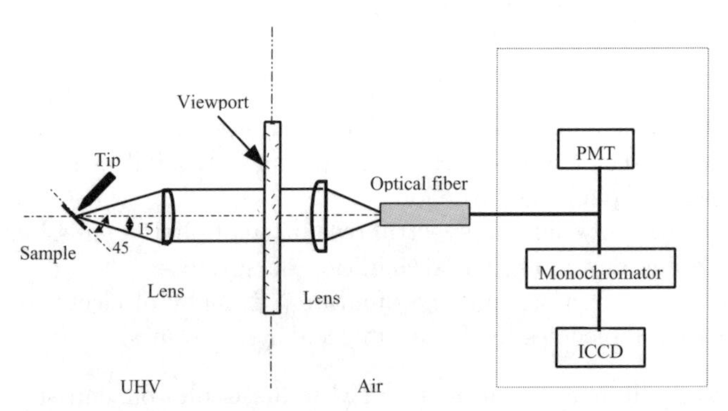

Fig. 2.3. Schematic diagram of our photon detection system through a double lens setup in combination with UHV–STM

Experiments were performed in an ultrahigh vacuum system (UHV) with facilities for sample heating and cooling, molecule deposition, and characterization by a scanning tunneling microscope (JEOL JSTM-4500XT). The base pressure of the system is less than 2×10^{-10} torr. Experiments were conducted at 80 and 295 K. Cu(100) substrates $3 \times 7 \times 0.5\,\mathrm{mm}^3$ were mechanically polished using $0.3\,\mu\mathrm{m}$ alumina paste followed by electrochemical polishing in a solution of 66% H_3PO_4 and 34% methanol (20 mA, 2.8 V, 10 min, 4°C). These were thoroughly rinsed in methanol before being loaded into the UHV chamber. Atomically clean Cu(100) surfaces were obtained after cycles of sputtering (Ar^+, 1 keV, $\sim 1\,\mu$A, $\sim$ 800°C, 30 min) and annealing (500°C, 30 min). Cu-tetra-(3,5 di-tertiary-butyl-phenyl)-porphyrin (Cu-TBPP) molecules were deposited onto Cu(100) by sublimation from a Knudsen cell at $\sim$ 300°C or higher. The sample was then either used as is or annealed at $\sim$ 200°C for 5 min. All images were taken in the constant-current topographic (CCT) mode with the sample biased. Tips used for imaging and photon emission were prepared by electrochemical etching of a W wire in a 2N NaOH solution followed by in situ cleaning in vacuum by Ar^+ ion sputtering or heating the tip apex up to $\sim$ 1 000°C.

A schematic view of our light collection and detection system is shown in Fig. 2.3. A condenser lens, 15 mm in diameter, is placed in UHV close to the tip–sample region (27.4 mm away) at an angle of 45° with respect to the surface. The collected solid angle is 0.22 sterad. Photons emitted from the tunneling gap are collected by this lens, transmitted through a viewport, and then refocused onto an optical fiber using another condenser lens located outside UHV. The fiber cable is then connected to either a PMT or a spectrophotometer (Hamamatsu PMA-100). The spectrometer includes a grating monochromator (C5094) and an intensified charge-coupled device (ICCD) with Peltier-cooled multialkali (S-20) and multichannel plate (MCP). The ICCD operates in a pulse counting mode at a dark count rate of ca. 20 counts per second (cps) at −15°C for the wavelength range 350–850 nm. To ensure that emitted photons are counted properly, the intensity was also measured by a photomultiplier (Hamamatsu H6180-01, dark counts $\sim$ 10 cps at 25°C, 300–650 nm), and a cooled digital CCD camera (Hamamatsu C4880-40, average dark counts 0.02 electron/pixel/sec at −50°C, 300–1100 nm).

2.2.2 Porphyrin Molecules

The electronic characteristic of metal-free porphyrin molecules is the 18 electrons delocalized along the inner 16-membered tetrapyrole ring of the molecule, forming a closed, completely conjugated aromatic system. This electronic structure is believed to be responsible for its optical behavior. Indeed, the optical absorption spectrum of porphyrins is successfully rationalized by Gouterman's four-orbital model [59]. This simple model gives rise to a relatively weak (pseudoparity-forbidden) Q-band in the visible region and an intense (optically allowed) Soret or B-band in the near-UV. These features are nicely reproduced in the optical absorption spectra of H_2-TBP porphyrin molecules with a fine-structured Q-band around 519.3 nm and a sharp B-band at 422.8 nm. The free-base TBP porphyrin molecules fluoresce strongly at 649 nm. Similarly for Cu-TBP porphyrins, absorption spectra show a small Q-band at 539.6 nm (2.3 eV) and a sharp B-band at 417.0 nm (3.0 eV), but they do not fluoresce at all. The ability of porphyrins to achieve various states of oxidation and the ease with which they bond to various metals has made them a major functional constituent of biological and molecular electronic systems. The introduction of metal atoms into the center of molecules enables us to tune not only oxidation states but also photophysical and photochemical properties of the molecules. A critical issue is how to understand their functions on the molecular level. To be more specific, what are the kinetic behavior and the mechanisms by which porphyrins oxidize, transfer energy, and relax? What is the effect of the central metal atom on the relaxation mechanism of electronically excited porphyrins [60]?

It is generally observed that metalloporphyrins with closed shell metal ions of Mg, Ca, and Zn exhibit fluorescence and phosphorescence, while open shell diamagnetic metal complexes of Ni do not emit at all; Pd complexes emit weak fluorescence and strong phosphorescence, while Pt complexes do not fluoresce, but do phosphoresce strongly. Paramagnetic metal complexes such as Cu, Ag, and Au either do not show any emission at all, or, as in copper complexes, exhibit short-lived phosphorescence [61]. Fe(II)-centered porphyrins relax to the ground state via nonradiative decay within 6 psec after excitation. In contrast, the metal-free porphyrin fluoresces strongly and its relaxation time to the ground state is longer than 1 nsec. Evidently, the central Fe(II) metal plays a predominant role in the electronic energy excitation of these porphyrins which proceeds via radiative or nonradiative decay channels.

Specifically for Cu(II)-porphyrins (oxidation states can be verified by ESR) [60], the interaction between the porphyrin electron system and the unpaired d electron of Cu(II) (d^9, $2S+1=2$) means that the ground state becomes a doublet $^2S_0$ ($2S+1=2$). The lowest excited singlet state becomes the lowest singdoublet $^2S_1$ that is closely related to the lowest excited singlet state of porphyrin. The lowest excited triplet states are tripdoublet $^2T_1$ and quartet $^4T_1$ ($2S+1=4$) originating from the lowest excited state of porphyrin. The singlet states of the metal-free porphyrin couple with the doublet levels of the central metal atom to form 'singdoublet' states ($^2S_1$). The triplet states couple with the doublets to form 'tripdoublets' ($^2T_1$) and quartet states ($^4T_1$). The coupling of singlet, doublet, and triplet states in Cu-centered TBP porphyrin molecules is responsible for the absence of fluorescence. A detailed kinetic investigation will help to clarify the decay pathways and the nature of transitions between different states.

In general, a good photoluminescence material is not necessarily a good electroluminescence material; furthermore, a bad photoluminescence material is usually not good for electroluminescence unless triplet states are used for photon emission.

2.2.3 Tunneling Transport Through Molecules and STM Imaging

An STM is a typical two-terminal device. Tunneling between two electrodes consists of elastic and inelastic portions with the former predominant. The inelastic tunneling channels contribute only a small fraction (less than 0.1%) to the total tunneling current and are affected by factors such as geometry, local dielectric properties, and density of states of both tip and surface [6]. The molecular image observed by STM is a combined result of topography and electronic structures of the molecule, tip, and substrate. To understand STM imaging from elastic tunneling and photon emission induced by inelastic processes, it is important to have a general picture of the transport behavior through individual molecules between two electrodes [3,62].

At low voltages, two regimes of elastic transport explain the large scatter of transparency values measured on single molecules. For a transmission coefficient larger than one ($T > 1$), the transport regime is ballistic: the molecular levels are in resonance with the Fermi level of the electrodes, and there is complete electronic delocalization along the molecule. For example, in the case of single-wall carbon nanotubes at low temperatures, the transport coherence length can extend to as much as 500 nm. Values of $T < 1$ indicate that there are some scattering events during electron tunneling: either the molecule is differently bound to the electrodes or the Fermi level is located within the gap between the highest occupied (HOMO) and the lowest unoccupied (LUMO) molecular orbitals. In the first case, the tunneling regime is determined by the metal–molecule contact, which can be optimized by controlling the surface chemistry. To overcome the second situation, the molecular orbitals need to be electronically coupled to the electrodes to stabilize a tunnel pathway for the carriers that is more efficient than vacuum. This pathway builds up from the constructive or destructive superposition of tunnel channels. For systems with $T < 1$ and involving a molecule symmetrically chemisorbed on the electrodes, such as C_{60} bound to two similar electrodes [63], two-terminal measurements exhibit linear I–V characteristics when V is much lower than the molecule's effective barrier height. This Simmons tunneling regime is analogous to electron tunneling between metals at low voltages, resulting in linear I–V curves. Nonlinear I–V curves usually occur at higher voltages in systems with $T > 1$ involving a physisorbed molecule, owing to the dominant effect from the contact tunnel barriers at the electrodes.

Tunneling through atoms and molecules whose filled and empty electronic states do not lie around the Fermi level occurs by a process known as virtual or off-resonance tunneling [3,64,65] at low voltages. The electrical resistance of the junction at contact is defined by the electronic transparency of the molecule. Transparency is a measure of the molecule's efficiency in extending the metallic wave functions of the electrodes. It is approximately proportional to the square of the inter-electrode electronic coupling introduced by the molecule with respect to the corresponding vacuum gap. Ohmic dissipation in the electrodes is one way to evaluate electronic transparency (T) from the macroscopically measurable quantity of tunnel resistance R, as $T = (h/2e^2)R^{-1}$, where e is the elementary charge and h is Planck's constant. However, owing to R being the resistance of the metal–molecule–metal tunnel junction, rather than that of the molecule itself, it cannot be used to define molecular conductivity.

The coupling of molecular electronic states with those of electrodes mainly occurs through the tails of the molecular orbitals, in particular those of the highest occupied and lowest unoccupied orbitals. These frontier orbitals provide such an extension through the induced shift of molecular orbital levels [66] and are sensitive to changes in hybridization and degeneracy induced by external distortion. For example, a mechanically induced distortion of 0.1 nm is sufficient to modulate the current flow through C_{60} by a factor of 100 [63]. Increasing this distortion leads to unit transparency, equivalent to a quantum unit of conductance $2e^2/h$. Molecular extension of the metallic wave functions is particularly relevant to devices operating at low voltages and where linear current–voltage characteristics are desirable. On the other hand, single-electron tunneling associated with Coulomb blockade effects has been observed for C_{60} molecules when they are deposited on an insulating layer at 4.2 K [67]. The wave function decoupling of molecules is suggestive of a single-electron device using a molecule as an ideal quantum dot [68].

The above description of transport through individual molecules suggests that access to the true molecular conductivity is a challenging issue. But by knowing a variety of factors involved in the transport process for a molecule sandwiched between two electrodes, one can have a better understanding of STM imaging, coupling of molecules with electrodes, molecular conformation on surfaces, and even the molecular manipulation process [69,70].

Figure 2.4 shows the structure and conformation of a Cu-TBPP molecule. The steric repulsion among the bulky substituents (legs) on the phenyl drives the phenyl rings roughly perpendicular to the porphyrin ring. STM images of a Cu-TBPP molecule on Cu(100) consist of four bright lobes, as shown in Fig. 2.5a for a very low coverage and Fig. 2.5b for a coverage close to one monolayer. The distance between the trans-lobes is $\sim$ 1.8 nm, slightly less than the molecular dimension of $\sim$ 2 nm. The bright lobes are thus assigned to the di-t-butylphenyl (DTP) side-groups, in accordance with previous reports [69,70] and theoretical simulation [71]. The characteristic four-lobe pattern is a registry of Cu-TBPP and is used for subsequent molecular recognition and controlled tip-positioning above a molecule.

The molecules are found to be relatively mobile on the surface at room temperature. As regards the stability of molecules against bias voltages and currents, our STM studies show that the molecules are stable below 3.8 V for a small tunneling current (e.g., 0.1 nA), but are easily broken into fragments when bias voltages are above 4.5 V. The application of a large current tends to drag the molecules across the surface during scanning. Porphyrin molecules are chosen because they are fluorescent molecules with aromaticity and are known to have high photoluminescence efficiency for converting π–π^* electronic excitations into visible and near-UV light in their free state or in solution [72]. However, in close proximity to metallic surfaces, nonradiative damping of molecular excited states by metal surfaces is thought to drastically quench fluorescence [52]. Our experiments on planar TPP-porphyrin molecules [with the $C(CH_3)$ replaced by H] appear to support the above claim. In order to suppress the fast electron energy dissipation to metal surfaces, a porphyrin architecture that is electronically decoupled from the substrate appears necessary. The use of bulky *t*-butyl groups makes the plane of phenyl rings rotate out of the plane of the porphyrin ring, and this dramatically weakens the π electron delocalization over the porphyrin–phenyl

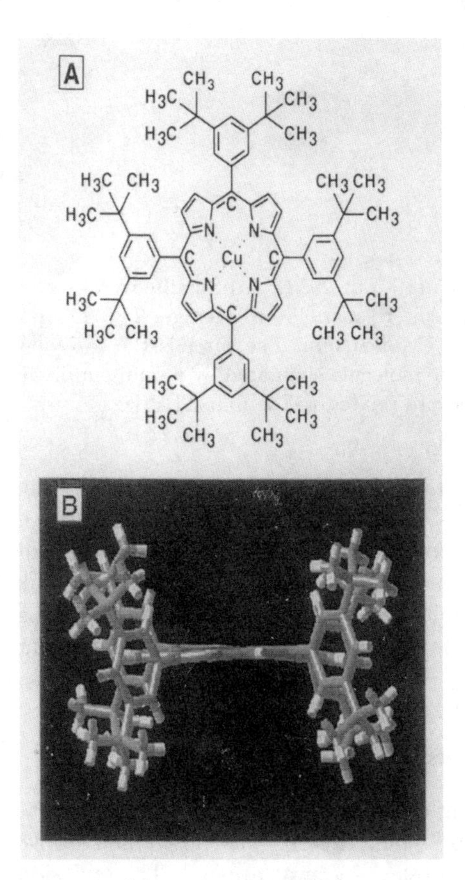

Fig. 2.4. (**a**) Structure and (**b**) conformation of Cu-TBP porphyrin molecules

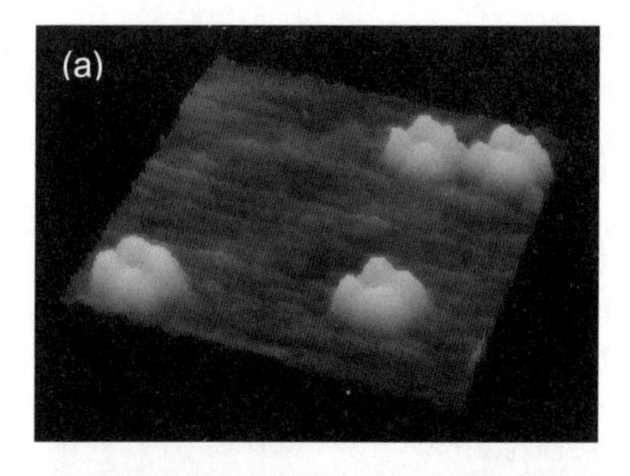

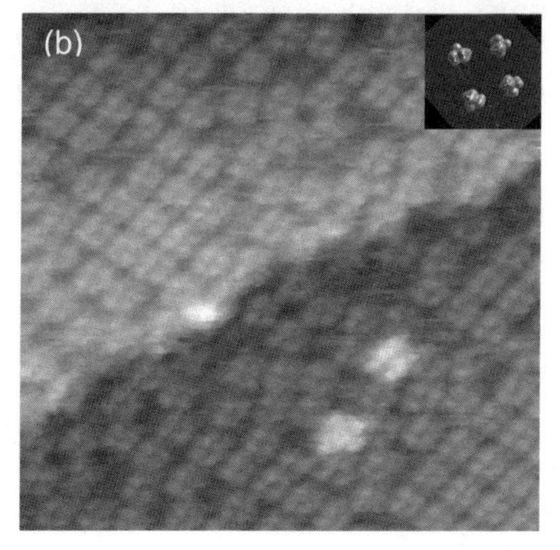

Fig. 2.5. STM images of Cu-TBPP on Cu(100) at 295 K: (**a**) Small coverage, $9 \times 9\,\text{nm}^2$, +0.5 V, 0.2 nA, no annealing applied; (**b**) monolayer coverage, $25 \times 25\,\text{nm}^2$, +1.5 V. 2 nA, sample annealed at $\sim 200°\text{C}$ for 10 min. The *inset* is the top view of the electronic density distribution of the molecule simulated by a semi-empirical INDO calculation. The features observed are related to the four TBP legs

bond. The electronic states of the porphyrin core are thus electronically decoupled from the surface since it hangs $\sim 5\,\text{Å}$ above it. Intense STM-induced photon emission is indeed observed from this molecule (see below). In addition, the crown-like cap configuration of Cu-TBPP together with its weak chemisorbed character on Cu(100) might serve as a squeezable tunneling junction and might allow the molecule to sustain a relatively large current for photon emission.

2.2.4 STM-Induced Photon Emission

Photon emission induced by tunneling electrons for the present system was initially detected by a CCD detector when the tip was positioned above molecules and a moderate bias voltage (e.g., 4 V) was applied. The emission event can occur for both 'on' and 'off' states of the feedback loop. The 'on' state gives reproducible and long-lasting photon emission (hours, but 'blinking') while 'off' state light emission lasts for only a few seconds, presumably due to the lack of a feedback loop to maintain a stable tunneling gap. Our intensity measurements indicate that light emission starts when the bias voltage is above 1.8 V in both polarities, and appears to reach a local maximum around 3.5–4 V. Further increase of the bias voltage up to 7 V (field-emission regime) often leads to short but incandescent emission ('bleaching'), which suggests that the molecule is modified or damaged, as revealed by subsequent STM imaging. On the other hand, the emission intensity increases almost linearly with current, as expected, and becomes visible to the naked eye at a current of around 50 nA. For a surface completely covered by Cu-TBPP molecules, since the tip is almost always above a molecule, light emission can be realized either when the tip is kept stationary above a molecule or during scanning. Variation of the bias from 2 to 5 V did not change the color of the emitted light, which appeared orange-red to the naked eye when the environment was relatively dark. It is worth pointing out that light is emitted only when the current fluctuates to some extent, indicating a relationship between photon emission and tunneling current behavior, although a quantitative relation is still hard to discern.

The first indication of enhanced light emission due to the molecules comes from our experiments performed on the surface in Fig. 2.5a with a very low coverage of Cu-TBPP molecules. Intense photon emission was readily detected by a CCD when the tip was positioned above a molecule, but little light was detected by CCD when the tip was located above the bare Cu surface under the same excitation conditions (e.g., 3.5 V, 10 nA); the latter was consistent with experiments performed on a clean Cu(100) surface. This observation agrees with reports concerning C_{60} on Au(110) [49] in which the detected photon intensity is greatest when the tip is placed above an individual C_{60} molecule. A molecular origin for light emission was also claimed previously by Flaxer et al. [53] in their study of 9,10-dichloroanthracene on an ITO substrate. A similar argument is proposed here based on the above observation and the optical spectral data given below. Since an STM tip is very sharp and generally believed to have only one atom at the tip apex, the enhanced emission above a single molecule also suggests that even single molecules can have a significant influence on the photon emission process.

2.2.5 Intensity and Quantum Efficiency

We are still pursuing efforts to obtain well-defined photon maps for molecules on surfaces at low tunneling currents through improvements in collection and detection systems. In this chapter, we report our preliminary results on emission intensity and optical spectra. Photon intensity and optical spectral data were acquired from a monolayer-covered surface in Fig. 2.5b instead of the one in Fig. 2.5a. This is because spectral data with a decent S/N ratio require a long integration time, but a low-coverage surface makes it difficult to hold the tip above the same molecule for a sufficient time due to the thermal drift problem. Although the intensity of photons emitted from an STM is generally low, the signal-to-noise ratio is actually quite high, above 50 in the present setup. Count rates of up to 10^4 cps per nA over 350–850 nm were recorded even for the small detection solid angle of 0.22 sterad. The total intensity was of the order of 10^8 cps for a tunneling current of around 10 nA after calibrating the quantum efficiency of the multialkali plate, MCP, solid angle, and attenuation or loss through transmission. (One can rule out the hypothesis that the variation in photon intensity arises from variations in tunneling currents by establishing that no direct structure correlation is observed in simultaneously recorded maps of the tunneling current.) The quantum efficiency of photon emission induced by STM for the present system was thus estimated to be around 10^{-4}–10^{-3} photons/electron. This value is about ten times larger than photon emission from pristine noble metal surfaces [6,28,58]. The enhancement of photon emission due to molecules is evident.

The so-called tip-induced plasmon (TIP) mechanism [20] alone, i.e., photon emission through inelastic tunneling processes with radiative decay of localized dipolar surface plasmons, cannot explain the enhanced emission above molecules. The presence of a molecule would lead to an increase in the distance between the tip and metal surface and thus to a smaller field enhancement (or TIP). This would instead result in intensity minima above molecules, contradicting the experimental observations. Conversely, the molecule plays an active role in the photon emission process and may even be an important photon source itself. Enhanced photon emission above molecules implies a strong coupling between the molecule and the electromagnetic modes of the cavity between the tip and metal surface [50]. In addition, the fluorescence of molecules may be enhanced in a way that is similar to that invoked in the classical explanation of surface-enhanced Raman scattering [51]. Another factor for the enhancement may pertain to the quantum confinement of electrons in the molecule [73] and the associated resonant tunneling process.

In principle, a direct electron–photon conversion is a forbidden process in the absence of a potential field because of their different dispersion relations. Surface plasmons on a perfectly smooth metal surface cannot radiate since such a process could not simultaneously conserve energy and momentum. However, when a tip is brought near the surface or when the surface is roughened, translational invariance along the surface is broken and photon emission can occur upon applying appropriate bias voltages. Although it has not yet been well understood how a molecule couples with the local electromagnetic field and enhances photon emission, it is reasonable to consider the molecule as a quantum system where tunneling electrons are more or less confined in the molecule during their transport. The confinement-induced enhancement in the radiative rate and reduction in the nonradiative rate would result in a higher efficiency of luminescence from the semiconducting molecule [73]. In a sense, each molecule on the surface can be viewed as a protrusion, which acts as a scattering center facilitating the conversion of plasmons into photons [74]. The unique crown-like conformation with bulky legs also helps to suppress the otherwise fast nonradiative decay [54] by weakening the interaction of the molecule with the substrate.

The model developed by Johansson et al. [29] takes the role of the tip into account by approximating the tunneling from a metallic sphere to a planar metal surface. The intensity of emitted light from tunneling electrons was found to split into two separate factors. One is the strength of the electromagnetic field in the vicinity of the tip giving us a measure of the coupling strength between tunneling electrons and electromagnetic field. The other is a fluctuation in the tunneling current that acts as a source for radiation. The presence of a tip close to a sample surface creates a localized interface plasmon built up from surface charges of opposite polarity on the tip and sample surface, respectively. This plasmon resonance leads to a considerable enhancement of the light emission from tunneling electrons.

Similar arguments involving current fluctuations may explain the enhancement effect for a surface with absorbed molecules. An electronically excited molecule can be viewed as an oscillating charge distribution or a dynamic dipole moment surrounded by an oscillating electric field. Molecules are very likely subject to structural deformation during tunneling excitation at high bias and current. This could cause a small change in the molecular or tunneling resistance and thus result in small fluctuations of tunneling currents even though a feedback circuit is used to maintain a constant current.

Two models have been proposed to address the excitation mechanism of molecules [57]. The first model views the tip–molecule–substrate configuration as a double barrier structure. When the molecular electronic states are aligned with the energy bands of the electrodes, molecules are directly excited by inelastic tunneling. This is similar to the ballistic regime mentioned above for transport through molecules [3,62]. The second model, a more favored one, considers the tip–molecule–substrate configuration as a single barrier structure. Inelastic tunneling excites localized electromagnetic modes between the tip and substrate, and then these IET–TIP modes excite the molecule indirectly via coupling with molecular electronic states. Since the photon emission is related to the dielectric functions of both tip and sample, the dielectric properties of the molecules inside the tunneling gap will probably affect both the IET–TIP modes and IET probability. The other factor for the enhancement is related to the molecular luminescence, which contributes directly to light emission (see below).

2.2.6 Optical Spectra

Further evidence for the enhancement effect and even fluorescence contributions to light emission from molecules is revealed in the optical spectra acquired over a monolayer of porphyrins on Cu(100). Figure 2.6 shows the

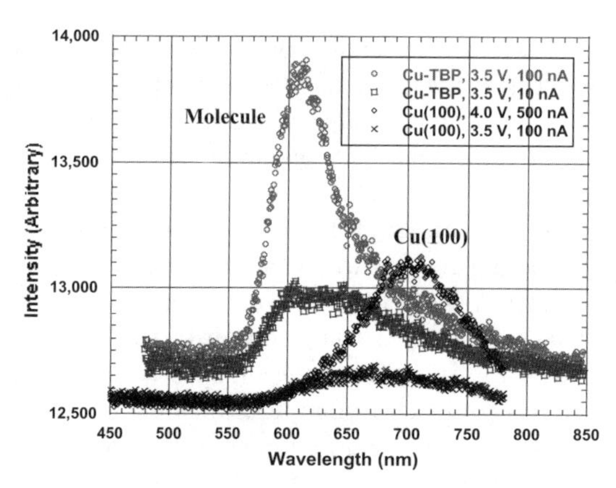

Fig. 2.6. Optical spectra induced by STM on Cu(100) at 295 K for pristine Cu(100) and Cu-TBPP molecules

typical optical spectra of Cu-TBP porphyrins induced by STM compared with those on the bare Cu(100) surface. The main peak occurs at $\sim$ 610 nm for a surface with porphyrin molecules, very different from the broad peak around 700 nm for the noble metal surface. Since the photon emission is related to the geometry of both tip and sample, peak positions may shift a bit in different experiments. A shifted optical spectrum is shown in Fig. 2.7, which was measured over 200 s from a different sample (monolayer coverage) and using a different tip. The excitation condition was set to +4 V and 100 nA. (It is worth noting that a large current was used in these experiments in order to get well-defined spectra, and this may complicate spectral analysis. The photon intensity in Figs. 2.6 and 2.7 is not corrected for the device response.)

In general, for bias voltages less than 2 V, no meaningful peaks were observed on the spectra even for a long integration time, suggesting a threshold around 2 V for light emission. The threshold at $V_b \sim 2$ V is in agreement with the HOMO–LUMO gap of Cu-TBPP molecules ($E_g \sim 2$ eV). Visible-light emission from molecules can occur only when $V_b \geq E_g$, no matter what kind of mechanism is involved, whether it be direct injection of electrons into the molecule or excitation of the molecule by coupling with local electromagnetic or plasmon modes. It is worth noting that adsorption of Cu-TBPP on Cu(100) results in a broadening of the HOMO and LUMO states, in particular the LUMO [70]. Accordingly, the tip can also contribute to enhancing the tunneling current by pressure-induced distortion of the molecular orbitals and resultant broadening when in contact with the molecule.

Figure 2.7 shows two broad peaks at 660 and 715 nm, respectively, both having contributions from the localized surface plasmons on Cu(100). Consistent with the spectra in Fig. 2.6, the peak at $\sim$ 715 nm is almost invariant with respect to the excitation voltage and sample location, suggesting an intrinsic feature associated with the Cu surface. However, the broad peak from 550 to

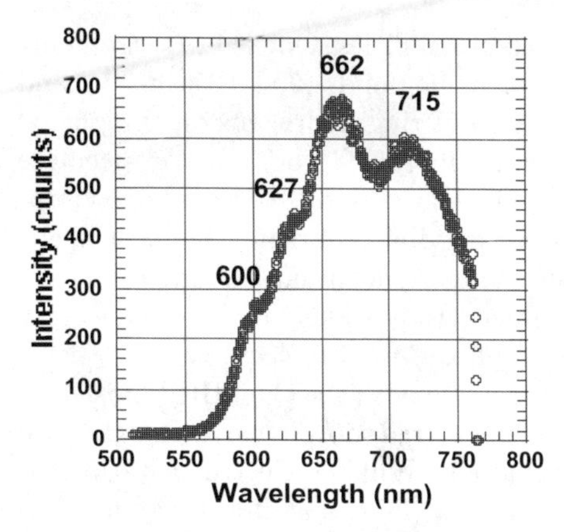

Fig. 2.7. Optical Spectra of Cu-TBPP on Cu(100) at 295 K at a preset excitation condition of +4 V and 100 nA integrated over 200 s

680 nm presents new features that are not observed for the pristine surface in terms of intensity enhancement and peak shape. This peak is therefore partly attributed to the molecules. Since Cu-TBP porphyrins do not fluoresce, the observed high-energy broad peak is compared with the fluorescence peak of the free-base H_2-TBP porphyrin molecules at 649 nm. The good agreement comes as no surprise because in any case fluorescence has the same origin as π–π^* electronic excitations from the porphyrin cores. There is nevertheless a small possibility that the Cu-centered porphyrins are modified to free-base porphyrins under the excitation condition for photon emission. The shape and intensity for this broad peak appear to be site-dependent, some with a shoulder-peak structure (Fig. 2.7), some with only a single peak (Fig. 2.6), implying that the spectra are related to the local molecular arrangements. The cutoff at 550 nm is attributed to the electronic interband transition in copper [20,75].

Another interesting feature in Fig. 2.7 is the presence of three shoulder peaks, at approximately 600, 627, and 662 nm, respectively, superimposed on the broad peak around 630 nm ($\sim$ 2 eV). It is possible that the fine structure is caused by the different cavity modes between the tip and sample. However, the peak spacing of the fine structure is $\sim 0.1\,\text{eV} \approx 800\,\text{cm}^{-1}$, which falls in the range of molecular vibrational transitions. This suggests another possible cause for the fine structure: the photon emission may arise from the coupling of molecular electronic states with the local electromagnetic modes. Excited molecules then undergo vibrational relaxation of excited states before decaying radiatively to the ground state. This speculation may be partly supported by the IR spectrum of the molecules, which does show a spectral peak around $800\,\text{cm}^{-1}$.

The attribution of this vibrational peak has not yet been clarified. However, referred to the range of vibrational frequencies reported for the chemical bonds of organic molecules, there are two possibilities. One is C–C stretching or bending between the phenyl-porphyrin single bonds (700–1250 cm^{-1}), and the other is the C–H deformation vibration of ring hydrogens on the phenyl groups or the ring deformation vibration itself (700–900 cm^{-1}). We speculate that the observed frequency at 800 cm^{-1} is related to one of the above vibrational modes, with the ring C–H deformation vibration being the most likely. A Franck–Condon-like (vertical transition) mechanism may be involved during the transition. The current or field fluctuation is likely to cause molecular deformation followed by the vibrational relaxation of molecules.

The observation of spectral peaks associated with the Cu-TBPP molecules is very exciting but puzzling, since Cu-centered porphyrin molecules are known not to fluoresce under photoexcitation. Will the electroexcitation inside a nanogap defined by STM bring out new physics about energy absorption, transfer, or decay processes? Does the mixing or coupling of singlet and triplet states for molecules behave differently in localized electromagnetic cavities? Or is it possible that the Cu-TBPP molecules lose their central metal atoms and are changed to free-base molecules when a high electrical field and large current are applied across the molecular junction? Further research is required to clarify these issues. It is worth noting that the broad peak associated with the molecules appears dramatically enhanced when the bias voltage is above 3 V, in good agreement with the high absorption efficiency of the B-band at 3.0 V for the optical excitation of Cu-TBP porphyrins. These observations suggest that, although the excitation or decay mechanism of molecules by electrons might be different from that by photons, the energy absorption behavior appears similar on the basis of the energy dependence of optical spectra.

2.3 Conclusion

We have demonstrated a technique to produce tunneling-electron-induced photon emission from Cu-TBP porphyrin molecules on a Cu(100) surface. The emission intensity and optical spectral data show not only the enhancement effects of the molecule, but also new features revealing the molecular origin of light emission. The envelope of the spectral peak associated with molecular fluorescence appears to exhibit a fine structure that is consistent with the vibrational transition of the molecule. The quantum efficiency is typically 10^{-4} photons per electron for Cu-TBP porphyrins. Since Cu-centered porphyrin molecules do not fluoresce under photoluminescence, the observation of unexpected light emission from the molecule may suggest new physics in the nanoscale STM excitation.

Although the data from the present work and others are still not sufficient to reach a definite conclusion about the nature of the observed phenomena, the general physical process of STM-induced molecular light emission can be pictured as follows. When an STM is positioned above a molecule and a bias voltage is applied across the tunneling gap, a net tunneling current flows from the tip to the substrate via the absorbed molecule. A tunneling electron or an absorbed molecule can excite a surface plasmon in the metal substrate, which causes an enhancement of the local electromagnetic field and in turn helps to excite the molecules. The electromagnetic excitation of the molecule occurs via the resonant coupling of molecular electronic states with the localized electromagnetic modes. The excited molecules relax vibrationally and then decay to the ground states with emission of light. The decoupling of luminescent porphyrin cores from the substrate, a means to suppress fast nonradiative energy dissipation, is found to be important for efficient light emission.

On the other hand, the combination of STM with optical techniques shows promise for carrying out electronic and vibrational spectroscopy on the single-molecule level. This hybrid technique will facilitate the identification of individual molecules ('molecular recognition') and luminescent defects on surfaces, permitting chemical mapping on the molecular scale. The intense photon emission from single molecules may be a potential point source of light that could be used for high resolution imaging and even for flat-display technology. The idea of having a variety of molecular chromophores in a component to yield specific responses upon excitation may be a way to address the input and output issue of electronic signals via localized electromagnetic modes. The use of plasmons from nanoclusters to amplify the gain in light transistors has recently been demonstrated for optical circuits [76] and similar concepts of plasmon enhancement may be useful for the design of molecular-scale amplifiers and transistors. In addition, an understanding of electron transport in the tip–molecule–substrate geometry may provide insights into some biological processes since electron transfer from electron-donor to electron-acceptor molecules via a molecular bridge is an important feature of many biological and chemical systems [77].

Acknowledgments

We thank Dr. J.K. Gimzewski and Dr. R. Berndt for helpful discussions. This work was supported by the Organized Research Combination System in Japan.

References

1. W.E. Moerner and M. Orrit: Science **283**, 1670 (1999)
2. J.K. Gimzewski: Photons and Local Probes, ed. by O. Marii and R. Moller (IBM, Netherlands 1995) p. 189
3. J.K. Gimzewski and C. Joachim: Science **283**, 1683 (1999)
4. R. Berndt: Scanning Probe Microscopy, ed. by R. Wiesendanger (Springer, Berlin 1998) Chap. 5, p. 97
5. B.C. Stipe, M.A. Rezaei, and W. Ho: Science **280**, 1732 (1998)

6. R. Berndt and J.K. Gimzewski: Phys. Rev. B **48**, 4746 (1993)
7. J. Lambe and S.L. McCarthy: Phys. Rev. Lett. **37**, 923 (1976)
8. J.K. Gimzewski, B. Reihl, J.H. Coombs, and R.R. Schlittler: Z. Phys. B **72**, 497 (1988)
9. J.K. Gimzewski, J.K. Sass, R.R. Schlittler, and J. Schott: Europhys. Lett. **8**, 435 (1989)
10. B. Reihl: Surf. Sci. **162**, 1 (1985)
11. D.L. Abraham, A. Veider, C. Schonenberger, H.P. Meier, D.J. Arent, and S.F. Alvarado: Appl. Phys. Lett. **56**, 1564 (1990)
12. P. Renaud and S.F. Alvarado: Phys. Rev. B **44**, 6340 (1991)
13. T. Tsuruoka, Y. Ohizumi, S. Ushioda, Y. Ohno, and H. Ohno: Appl. Phys. Lett. **73**, 1544 (1998)
14. R. Berndt and J.K. Gimzewski: Phys. Rev. B **45**, 14095 (1991)
15. S.F. Alvarado and P. Renaud: Phys. Rev. Lett. **68**, 1387 (1992)
16. S. Ushioda: Solid State Commun. **84**, 173 (1992)
17. S. Sasaki and T. Murashita: Jpn. J. Appl. Phys. **38**, L4 (1999)
18. A. Downes and M.E. Welland: Phys. Rev. Lett. **81**, 1857 (1998)
19. C. Thirstrup, M. Sakurai, K. Stokbro, and M. Aono: Phys. Rev. Lett. **82**, 1241 (1999)
20. R. Berndt, J.K. Gimzewski, and P. Johansson: Phys. Rev. Lett. **67**, 3796 (1991)
21. J.H. Coombs, J.K. Gimzewski, B. Reihl, J.K. Sass, and R.R. Schlittler: J. Microsc. **152**, 325 (1988)
22. R. Berndt and J.K. Gimzewski: Ann. Physik. **2**, 133 (1993)
23. A. Downes, M.E. Taylor, and M.E. Welland: Phys. Rev. B **57**, 6707 (1998)
24. D. Hone, B. Muhlschlegel, and D.J. Scalapino: Appl. Phys. Lett. **33**, 203 (1978)
25. R.W. Rendell, D.J. Scalapino, and B. Muhlschlegel: Phys. Rev. Lett. **41**, 1746 (1978)
26. J.R. Kirtley, T.N. Theis, J.C. Tsang, and D.J. MiMaria: Phys. Rev. B **27**, 4601 (1983)
27. R. Berndt, J.K. Gimzewski, and P. Johansson: Phys. Rev. Lett. **71**, 3493 (1993)
28. B.N.J. Persson and A. Baratoff: Phys. Rev. Lett. **68**, 3224 (1992)
29. P. Johansson, R. Monreal, P. Apell: Phys. Rev. B **42**, 9210 (1990)
30. K.R. Welford and J.R. Sambles: J. Mod. Opt. **35**, 1467 (1988)
31. Y. Suzuki, H. Minoda, and N. Yamamoto: Surf. Sci. **438**, 297 (1999)
32. R. Berndt and J.K. Gimzewski: Surf. Sci. **269/270**, 556 (1992)
33. M. Tsukada, T. Schimizu, and K. Kobayashi: Ultramicroscopy **42–44**, 360 (1992)
34. P. Johansson: Ph.D. Thesis (Physics Department, Chalmers University, Goteborg, Sweden 1991)
35. A.G. Malshukov: Phys. Rep. **194**, 343 (1990)
36. R. Berndt, R. Gaisch, J.K. Gimzewski, B. Reihl, R.R. Schlittler, W.D. Schneider, M. Tschudy: Phys. Rev. Lett. **74**, 102 (1995)
37. Y. Uehara, Y. Kimura, S. Ushioda, and K. Takeuchi: Jpn. J. Appl. Phys. **31**, 2465 (1992)
38. A.L. Vasquez de Parga and S.F. Alvarado: Phys. Rev. Lett. **72**, 3726 (1994)
39. N. Majlis, A.L. Yeyati, F. Flores, and R. Monreal: Phys. Rev. B **52**, 12505 (1995)
40. P. Johansson: Phys. Rev. B **58**, 10823 (1998)
41. K. Takeuchi, Y. Uehara, S. Ushioda, S. Morita: J. Vac. Sci. Technol. B **9**, 557 (1991)

42. Y. Uehara, T. Fujita, and S. Ushioda, K. Takeuchi: Phys. Rev. Lett. **83**, 2445 (1999)
43. I. Smolyaninov: Surf. Sci. **364**, 79 (1996)
44. M. Xiao: Phys. Rev. Lett. **82**, 1875 (1999)
45. P. Johansson, R. Berndt, J. Gimzewski, and S. Apell: Phys. Rev. Lett. **84**, 2034 (2000)
46. M. Xiao: Phys. Rev. Lett. **84**, 2035 (2000)
47. P. Chaumet and A. Rahmani: Phys. Rev. Lett. **84**, 3498 (2000)
48. G.S. Agarwal: Phys. Rev. A **11**, 230 (1975); **12**, 1475 (1975)
49. R. Berndt, R. Gaisch, J.K. Gimzewski, B. Reihl, R.R. Schlittler, W.D. Schneider, M. Tschudy: Science **262**, 1425 (1993)
50. A. Adams, J. Moreland, and P.K. Hansma: Surf. Sci. **111**, 351 (1981)
51. A. Otto, I. Mrozek, H. Grabhorn, and W. Akemann: J. Phys. Condens. Matter **4**, 1143 (1992)
52. D.H. Waldeck, A.P. Alivisatos, and C.B. Harris: Surf. Sci. **158**, 103 (1985)
53. E. Flaxer, O. Sneh, and O. Cheshnovsky: Science **262**, 2012 (1993)
54. P. Avouris and B.N.J. Persson: J. Phys. Chem. **88**, 837 (1984)
55. V. Sivel, R. Coratger, F. Ajustron, J. Beauvillain: Phys. Rev. B **45**, 8634 (1992)
56. D. Fujita, T. Ohgi, W.-L. Deng, H. Nejo, T. Okamoto, S. Yokoyama, K. Kamikado, and S. Mashiko: Surf. Sci. **454–456**, 1021 (2000)
57. V.J. Langlais, R.R. Schlitter, and J.K. Gimzewski: private communication
58. R. Berndt, R.R. Schlitter, and J.K. Gimzewski: J. Vac. Sci. Technol. B **9**, 573 (1991)
59. M. Goutterman: J. Mol. Spectrosc. **44**, 37 (1972)
60. T. Kobayashi, D. Huppert, K.D. Straub, and P.M. Rentzepis: J. Chem. Phys. **70**, 1720 (1979)
61. F.R. Hopf and D.G. Whitten: Porphyrins and Metalloporphyrins, ed. by K.M. Smith (Elsevier, New York 1972) p. 667
62. C. Joachim, J.K. Gimzewski, and A. Aviram: Nature **408**, 541 (2000)
63. C. Joachim, J.K. Gimzewski, R.R. Schlittler, and C. Chavy: Phys. Rev. Lett. **74**, 2102 (1995)
64. C. Joachim and J.K. Gimzewski: Europhys. Lett. **30**, 409 (1995)
65. A. Yazdani, D.M. Eigler, and N. Lang: Science **272**, 1921 (1996)
66. M. Magoga and C. Joachim: Phys. Rev. B **56**, 4722 (1997)
67. D. Porath and O. Milo: J. Appl. Phys. **81**, 2241 (1997)
68. M.A. Reed, C. Zhou, C.J. Muller, T.P. Burgin, and J.M. Tour: Science **278**, 252 (1997)
69. T.A. Jung, R.R. Schlittler, and J.K. Gimzewski: Nature **386**, 696 (1997)
70. J.K. Gimzewski, T.A. Jung, M.T. Cuberes, and R.R. Schlittler: Surf. Sci. **386**, 101 (1997)
71. P. Sautet and C. Joachim: Chem. Phys. Lett. **185**, 23 (1991)
72. I.B. Berlam: Handbook of Fluorescence Spectra of Aromatic Molecules (Academic Press, New York 1965)
73. A.P. Alivisatos: Science **271**, 933 (1996)
74. N. Kroo, J.P. Thost, M. Volker, W. Krieger, and H. Walther: Europhys. Lett. **15**, 289 (1991)
75. M. Welkowsky and R. Braunstein: Solid State Commun. **9**, 2139 (1971)
76. J. Tominaga, C. Mihalcea, D. Buchel, H. Fukuda, T. Nakano, N. Atoda, H. Fuji, and T. Kikukawa: Appl. Phys. Lett. **78**, 2417 (2001)
77. W.B. Davis, W.A. Svec, M.A. Ratner, and M.R. Wasielewski: Nature **396**, 60 (1998)

3 Photon Counting Methods in STM and SMS

A.G. Vitukhnovsky and I.S. Osad'ko

3.1 Introduction

Combined electrical and optical measurement methods applied to a sample are able to supply us with comprehensive information concerning condensed matter. However, up to now, such a combination has only been carried out for semiconductors. Optical methods are ineffective in metals, whereas methods based on conductivity are ineffective in organic systems, which are generally insulators. The invention of tunneling microscopy [1,2] opened up new possibilities for combining electrical and optical methods because scanning tunneling microscopy (STM) enables us to use electric currents to excite single organic molecules.

Up to now the main sources of information concerning physical characteristics of small junctions in STM are DC current–voltage curves for these junctions [3]. It should be stressed that STM provides the best space resolution attainable by physical methods today. In fact, we can see a single atom using STM [4]. As far as the time resolution of scanning tunneling microscopy is concerned, it is not particularly high and can only distinguish on the millisecond scale. STM does not 'see' any faster dynamics. Consequently, STM displays an averaged picture when faster relaxation processes exist in a junction.

On the other hand, impressive progress has been made in single molecule spectroscopy (SMS) over the last decade [5–9]. SMS uses an excitation fluorescence method that was widely employed before in the traditional spectroscopy of molecular ensembles to measure absorption spectra. However, a laser source of light can provide space resolution comparable with the wavelength. Therefore, we have to use very low concentrations of guest molecules in SMS. When the distance between individual guest molecules becomes large we are able to excite a single molecule. By using near-field scanning optical microscopy (NSOM), we can raise the space resolution of the setup, although it remains inferior to that reached by STM [10].

However, SMS has some advantages over STM. Firstly, the time resolution of SMS can reach the nanosecond level [11]. Secondly, SMS can probe guest molecules other than those lying on the surface. The third advantage stems from the possibility of using monochromatic laser excitation because the frequency of laser excitation is an additional external parameter that can easily be changed in an experiment.

The majority of the reviews in this book discuss problems of electric current in STM. We therefore decided to concentrate our discussion on problems concerning photons emitted in STM or registered in SMS with an emphasis on SMS methods. We intend to discuss in detail theory, experimental data and possible applications of SMS because this information may be useful in applying photon counting methods in STM and interpreting experimental data when STM probes organic molecules on surfaces.

3.2 Light Emission from STM

Let us start with a short review of a problem which concerns light emission from small tunnel junctions. A schematic view of STM is shown in Fig. 3.1.

If a bias of a few volts is applied between the metallic tip and surface, an electric current from the tip to the surface emerges when the distance d between the tip and the surface falls to nanometric scales. The current is due to a single-electron transition through the potential barrier between the tip and the surface. The tunneling nature of the current is confirmed by its strong exponential dependence on the distance d. The tunneling current I is increased if the bias V is raised.

The $I(V)$ dependence shown in Fig. 3.2 is the main physical characteristic of the junction in STM. This is a nonlinear function. It reveals a breakdown of Ohm's law because the junction conductivity depends on the bias.

The bias-dependence of conductivity sometimes exhibits non-monotonic behavior, as can be seen in Fig. 3.2. The conductivity is proportional to the rate of single-electron tunneling through the potential barrier in the junction. As has been demonstrated by direct numerical calculation in a double-well potential model [13,14], the tunneling transition probability is increased by one order of magnitude if there is a resonance between levels in the left- and right-hand wells separated by a potential barrier. The oscillations observed

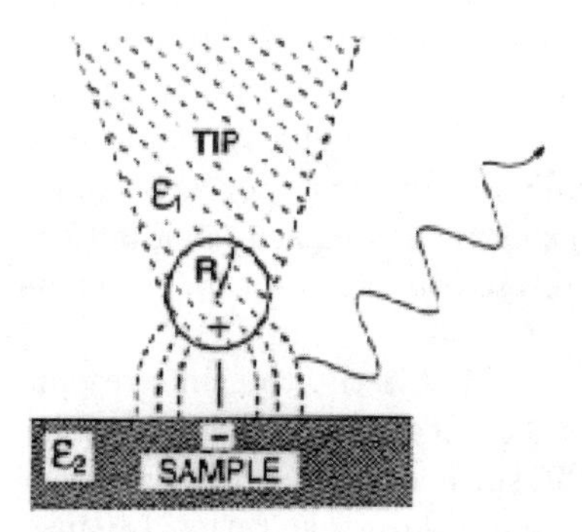

Fig. 3.1. Tip–surface tunneling junction in STM [12]

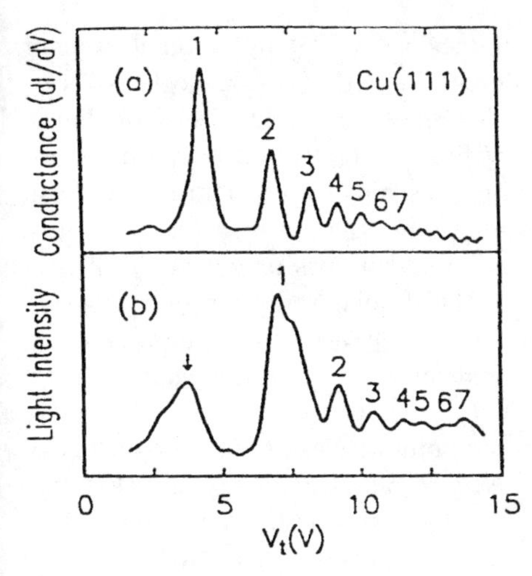

Fig. 3.2. Conductance of the tunneling junction between W tip and Cu(111) surface vs. bias [12]

in this rate and shown in Fig. 3.2 can be ascribed to this resonant effect. This means that single-electron tunneling is accompanied by the creation of electromagnetic field quanta in accordance with the following energy balance

$$eV = n\hbar\Omega \,, \tag{3.1}$$

where the left-hand side describes the change in the tunneling electron energy and n is the number of quanta created by the tunneling electron.

Due to its high space resolution, STM is a very promising tool for probing single molecules on a surface. Various guest molecules have been investigated using tunneling current excitation, including C_{60} [15–17], porphyrin derivatives [18,19] and molecular complexes [20]. Fluorescence excited by tunneling electrons was the main source of information concerning these molecules.

However, there is a general problem in this type of spectroscopic experiment. The tip–surface tunneling junction can emit light even if there are no guest molecules in the junction. Transformation of electronic energy into light energy is detected in junctions which do not include any guest molecules [21,22]. How can we separate molecular fluorescence from light emitted by the junction itself? What is the mechanism for transforming electronic energy into light?

There are various points of view concerning mechanisms for this energy transformation. From the point of view of classical electrodynamics, a constant bias creates a constant electric current. In this case only radiation of the Cherenkov type is possible [23]. This idea is discussed in [24]. However, it is unlikely that tunneling currents can be considered in this classical fashion.

Johansson et al. [25] used formulae for quantum transitions in order to take the tunneling current into account. The spectral shape of light emission was calculated in [22] with the help of the theory in [25] and good agreement with experimental data was found. Unfortunately, this theory neglects retardation effects in the electromagnetic field. Smolyaninov [24] believes that it is impossible to neglect retardation and that the agreement between experimental data and calculations declared in [22] seems to be fortuitous. Indeed, if we neglect the time t_0 which determines the difference in retardation of the electromagnetic wave emitted by different charges or currents, all possible radiation disappears. For instance, the dipolar radiation only emerges if, expanding electromagnetic potentials in power series in the parameter t_0, we go to first order ωt_0 in the approximation, where ω is the characteristic frequency of the classical current.

At the present time the most plausible point of view is that the energy of a single tunneling electron is transformed into plasmon energy and that it is plasmons which then create photons [12]. However, the theory based on this idea in [12,22,25] is not only questioned by Smolyaniniv [20]. Xiao [27] also criticizes the approach developed in [12,22,25], arguing against any important role for plasmons in the photon emission process. Xiao tries to explain light emission from the tip–surface junction using a simplified Hamiltonian

$$H = H_0 + eV \cos \omega t \,, \tag{3.2}$$

where H_0 is the Hamiltonian of the electron moving through the potential barrier between the tip and surface and V is the applied bias. It is claimed that a system with this simplified Hamiltonian is able to describe emission from the tip–surface junction and that plasmons are not needed to explain emission. The arguments of both groups can be found in letters [28,29]. Even this brief review of approaches to the problem of light emission from the tip–surface junction clearly shows that the theory of such emission is far from settled.

Since tunneling transitions are of a quantum nature, they occur randomly in time. Hence emitted photons accompanying this tunneling will likewise be distributed randomly in time. Put another way, time intervals between successively emitted photons will fluctuate. At a current of 1 nA, the average time between successively emitted photons is of the order of 10^{-10} s. A PMT will reliably detect a train of photons even if the quantum efficiency of detection is about 10^{-5}. The distribution of time intervals between emitted photons contains information about the dynamics of the tunneling junction. To our knowledge, this information concerning STM junctions has not yet been taken into consideration. In SMS, on the contrary, the temporal behavior of fluorescence measured in the photon counting regime is the main source of information about single-molecule dynamics. One- and two-photon counting methods in SMS serve as reliable sources of information concerning the dynamics of single molecules interacting with their environment [30,31]. The main goal of this review is to discuss achievements using these methods in the context of possible applications to scanning tunneling microscopy.

3.3 Role of Photon Counting in Time-Resolved Spectroscopy

Over the last two decades the methods of photon counting have spread from the narrow confines of nuclear physics applications to the much broader area of common spectroscopy. The new generation of photomultipliers (PMTs) with microchannel plates and avalanche diodes has given additional impetus for applications of photon counting methods to far- and near-field optics. Photon counting has important advantages over traditional measurement methods for weak optical signals, such as modulated spectroscopy (e.g., the lock-in amplifier method and other analog photocurrent measurements).

Upon photo-excitation an organic molecule is promoted to a non-equilibrium state from which it relaxes back to the ground state. Time-resolved photoluminescence spectroscopy is the method of choice for obtaining information on relaxation pathway(s). If the primary excitation event creates a vibrationally hot singlet (S_1) state or a higher-lying S_n state, then relaxation can in principle involve vibrational cooling, internal conversion, intersystem crossing, readjustment of molecular selection, and finally, radiative or non-radiative decay to the ground state. In dense media, solvation effects as well as energy transfer and dissociation into a pair of charge carriers can also occur. Which of these processes can be monitored by time-resolved spectroscopy depends on the time resolution of the experiment.

Conventional methods with nanosecond time resolution yield the lifetime of an emitting state, whilst the preceding history of that state remains concealed, however. The advent of picosecond (ps) and femtosecond (fs) laser systems as excitation sources and of streak camera and up-conversion techniques on the detection side allow extension of the time resolution of fluorescence spectroscopy into the sub-picosecond regime. This has opened the way to studying ultra-fast photophysical relaxation processes such as the solvation of an excited chromophore in solution, dephasing of optical excitations, and energy transfer in condensed systems.

Photon counting makes it possible to apply statistical methods for processing measurement results. It means that we can measure kinetic processes in optics using pulse trains rather than single shots. Well known achievements of picosecond and femtosecond laser techniques give impressive examples of the high repetition of short laser pulses (less than 10 fs). The modern technique of time-correlated single photon counting gives physical resolution less than 20 ps, but this value can be improved by an order of magnitude by employing special deconvolution methods.

3.4 Ensembles and Single-Molecule Spectroscopy Studies

The huge progress in semiconductor nanophysics has led to studies of tiny molecular structures right down to single molecules. Electrical and optical properties of molecular nanostructures are under investigation around the world. Of course, the study of fast processes for electron system relaxation in restricted geometries is a key point. Effective combination of methods with high spatial and time resolution can achieve the desired result.

Two main approaches are used in this field – confocal and near-field scanning optical microscopy. Both methods can combine with time-resolved spectroscopy based on single photon counting. The main advantage of confocal microscopy is large value of optical signal (number of emitting photons) in contrast with NSOM approach. In case of NSOM the best spatial NSOM resolution (better that 20 nm) can be reached and the optical signal weakness can be overcome by increasing the time of experiments for good photon statistics assurance.

In practice, it is often necessary to identify different analytes in a given sample. The authors of [32] reported measurements of spectroscopic properties in a matrix at the single-molecule level. In these experiments, differences in the spectral emission properties of individual molecules of Rhodamine 6G and Texas Red in combination with two separate detectors were used to distinguish the two dyes.

To distinguish different fluorescent dyes or to identify tagged analyte molecules, it is not necessary to collect nearly as many photons as are needed for an exact lifetime measurement with the time-correlated single-photon counting (TCSPC) technique. The problem then consists in comparing raw data with certain well-defined fluorescence decay patterns measured with high precision. Hence, from a statistical point of view, one is confronted with a classification problem. An optimal classification method has already been developed in the framework of information theory and is described in detail elsewhere [33–36].

An important question arises for the time-resolved identification of single dye molecules in a matrix: how many photons are necessary to distinguish two or more dye labels with different lifetimes? A first approximation for the probability of misclassification can be obtained by Cramer's inequality [36]. Hence, in combination with an efficient maximum likelihood estimator (MLE) algorithm (3.4) [37], the identification of different dye molecules on the single-molecule level should be possible due to their characteristic fluorescence lifetimes.

Zander et al. [38] first demonstrated the identification of single rhodamine dye molecules in water by their characteristic fluorescence lifetimes of 1.79 ± 0.33 ns (Rhodamine B) and 3.79 ± 0.38 ns (Rhodamine 6G) excited by a frequency-doubled titanium/sapphire laser emitting at 514 nm. This was achieved using confocal microscopy in combination with the technique of time-correlated single-photon counting. Using an efficient modified MLE algorithm, the classification probability was $> 90\%$ with 100 collected photons per burst.

The detection volume is usually defined by the intersection of the focused excitation laser and the image of a spatial filter. Reduction of the detection volume improves the signal-to-background ratio by many orders of magnitude without measurable photodestruction of the dye molecules under study. Poisson statistics predicts, for a concentration of less than 10^{-10} M, that the number of molecules fluctuates predominantly between 0 and 1 in an applied detection volume in the femtoliter region. Hence, the probability of two or more molecules being present in the detection volume simultaneously is negligible at concentrations below 10^{-10} M.

On the other hand, a small detection volume reduces the transition time, i.e., the measurement time. During the transition through the detection volume, the dye molecule is excited by the laser light from the ground state S_0 into high-lying vibrational levels of the first excited state S_1, which undergoes rapid non-radiative internal conversion to low-lying S_1 levels. The optical saturation limit is the maximum rate that a dye molecule can be cycled between S_0 and S_1 and is dependent on the fluorescence lifetime τ_f of the dye.

Besides irreversible photodestruction, several depopulation pathways, such as intersystem crossing into the triplet state, compete with fluorescence emission thereby reducing the number of emitted photons. Since no photon is obtainable during the triplet lifetime of the excited dye, the triplet quantum yield and lifetime is a very important photophysical constant for single-molecule experiments. In other words, only those fluorescent dyes which exhibit very small triplet yields and short triplet lifetimes are suitable for single-molecule experiments. Consequently, most measurements on the single-molecule level in a matrix have been performed with rhodamine dyes. Additionally, rhodamine dyes emit up to 10^5 photons before photodestruction, even in aqueous solution [39].

To sum up, in order to identify dye molecules in a matrix, both spectral and time-resolved detection methods can be used. A detailed description of modern optical studies of single molecules can be found in other sections of this chapter.

3.5 Importance of Ensembles for Applications. Novel Ideas with Nanosize Ensembles

There is considerable scientific and technological interest in design, synthesis and characterization of organic materials that possess unusual solid-state properties. Of practical interest are nanosize molecular ensembles such as linear molecular aggregates, regular and non-regular dendrimers and other organic materials with restricted geometry in which cooperative effects have been observed.

The interest to exciton dynamics in dendritic polymers (also known as Cayley tree or Bethe lattices) and molecular agregates (J- and H-, or mixed type) arises from their peculiar dimensionality [41] and can be satisfied by complex steady state and time resolved optical experiments. The dynamics of electronic excitations in the fractal dimensional systems (restricted or unlimited), the direct exciton transport in hierarchically built macromolecules [42], the interaction of these large molecules to each other are some of the fundamental problems to be modeled. Highly symmetric and perfectly hyperbranched macromolecules with well controlled structures are good candidates for real exciton funnels. The approach [43] considers the high-branched wedges of dendritic molecules as light harvesting antennae, which pass the absorbed energy to initial core of the molecule playing a role of emitter. It is reasonable to suggest that such materials may be applied as a emitting layer of light emitting diodes (LEDs).

Most of the polymers reported for electroluminescence (EL) applications have linear structures and little attention has been paid to using three-dimensional polymers as the active emitters in LEDs [43]. Compared to linear polymers with similar molecular weights, three-dimensional hyperbranched polymers have lower intrinsic viscosity [44,45] and better solubility. However, the EL properties of hyperbranched polymers have not been reported. In considering the high functionality and desirable physical properties, combined with the robustness and ease of preparation of hyperbranched polymers, EL devices made from three-dimensional polymers are likely to have improved stability and efficiency compared with those made from general linear polymers. We try to extend two novel findings in this field: the light-harvesting effect and segregation of light-emitting Ln cations by dendritic shells. Both processes provide a way of enhancing luminescence by vectorial exciton migration to the core and isolation of luminescent core molecules from other such molecules. Applications of nanosize molecular ensembles (dendrimers or aggregates) can thus provide some advantages for practical applications.

By this approach, new features have recently suggested a method for electroluminescence study of thin films by applying a voltage between metal-covered NSOM tip and molecules/ensembles on a conductive substrate. Once again, photon counting can provide a wealth of information about the optical behavior of molecules/ensembles excited by tunneling currents.

3.6 Nanoensembles

3.6.1 Regular Dendrimers

As a new class of materials, dendrimers or 'starburst' polymers have generated great interest throughout the scientific community. Despite synthesizing

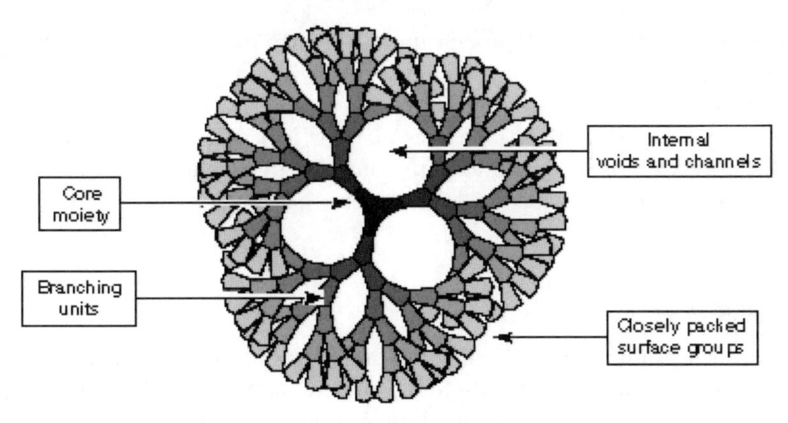

Fig. 3.3. Construction of typical regular dendrimer

difficulties, a wide range of these new materials have been produced and characterized [46].

However, the study of their properties is still in its infancy (compared with classical polymers) and there remains a large gap in our knowledge which must be filled if the potential uses of these new materials are to be realized. Dendrimers are highly branched structures, produced either by divergent or convergent synthetic methodologies which allow the formation of monodispersed nanoparticles of controlled dimensions and specific surface fictionalization.

The rationale for our studies on starburst dendrimers was based on the view that restrictions on the scale of single molecules of dimensions of the order 100 nm may result in novel electronic or optical phenomena not seen at other dimensions. For example, some correlated electronic effects on the dendrimer 'surface' at limiting generation would be observed such as fluorescence quenching or excimer formation.

3.6.2 Hyperbranched Polymers

Two distinct approaches to the synthesis of these hyperbranched polymers have evolved from the seminal work of Vogtle [47] and Denkewaller [48]. Tomalia's 'dendrimers' [46] and Newkome's 'arborols' [49] were independently developed through a process termed divergent synthesis. This approach is characterized by polymer growth emanating from a central core via an iterative protection/deprotection reaction scheme. Polymer growth typically begins from a 'core' molecule which undergoes exhaustive reactions with complementary monomers having two or more protected branch sites. Removal of the protecting groups and subsequent reaction of the liberated reactive sites leads to first-generation polymers. Repetition of this reaction process leads to polymers of the desired molecular weight, molecular size and topology.

The convergent synthetic approach independently developed by Hawker and Frechet [50,51] begins at what will eventually become the outer surface of the dendrimer. Polymer 'wedges' are synthesized via sequential reactions and contain single reactive functionalities at their loci. Wedges are then attached to a polyfunctional central core to complete dendrimer formation. A rapidly increasing number of cascade molecules has been synthesized and investigated.

3.6.3 J-aggregates

J-aggregates of cyanine dye molecules are nanostructures with sizes intermediate between molecular crystals and isolated molecules. Their optical properties are distinctly different from those of their monomeric constituents [52,53]. Interactions between the molecules in J-aggregates give rise to delocalized electronic excitations (Frenkel excitons), which are associated with a collective effect in the optical response [54–58].

Since they are delocalized over several monomer units, the excitations in J-aggregates are thought in turn to migrate over large distances (thousands of monomers) within the J-aggregates by an incoherent transfer mechanism [59]. Although these exciton transport properties in J-aggregates are rather important for applications, they have been understood to a much lesser extent than those of molecular crystals [60]. The same is also true for the geometry of the molecular arrangement within the J-aggregate. Regular brickwork [61], twisted brickwork [62] and several kinds of herringbone arrangements [63,64] have been proposed for J-aggregates. The unit cells of J-aggregate nanocrystals were assumed to contain one [65–67], two [68,69] or four molecules [70]. Moreover, different samples of J-aggregates consisting of the same dye molecules can contain a different number of molecules per unit cell depending on the preparation procedure [71,72].

3.7 Time-Correlated Single Photon Counting (TCSPC) Method

The marriage of fluorescence microscopy with time-correlated single photon counting has contributed to our understanding of nanosize ensembles.

Currently, most fluorescence microscopic imaging is performed as a measurement of emission or excitation intensity. These types of measurements have several limitations:

- intensity-based fluorescence imaging can be difficult to quantify,
- fast dynamic events (of the order of nanosecond to picosecond) in nanosize molecular systems cannot be studied,
- the emitted fluorescence from such samples is often complex and represents contributions from a number of molecular species which cannot be individually analyzed,

- autofluorescence and background fluorescence limit detection sensitivity.

In contrast, the measurement of fluorescence lifetimes does not suffer from these limitations. Fluorescence lifetimes possess the added benefits of being independent of local intensity, concentration and photobleaching of the fluorophore. Fluorophores with similar spectra may exhibit significant differences in their lifetimes, and the same fluorophore may display distinct lifetimes in different environments. Because fluorescence lifetimes are not affected by scattering, measurements of fluorescence lifetimes provide more sensitive and quantitative information about complex structures.

In most cases, periodic excitation pulses are used with a duty factor such that the luminescence between pulses manages to decay almost completely. Luminescence decay can be observed using an oscilloscope, either conventional or stroboscopic. Flash lamps or special discharges may be used as an excitation source.

An important variant of pulsed methods is the statistical method of photon counting. In this method, excitation is carried out with short pulses (1 ns or shorter) repeating at a frequency of 10 kHz or higher. Luminescence is detected by a photomultiplier operating in the photon counting mode. The luminescence intensity incident on the photocathode of the photomultiplier is weakened to such an extent that at most one photoelectron is knocked free from the photocathode after every excitation pulse. The time period between the excitation pulse and the moment of emission of this photoelectron is governed by the probability of light emission. The number of excitation pulses, i.e., those in which the first photoelectron they release has been recorded, is considerably smaller than the total number of pulses, and the mean number of these first photoelectrons as a function of time reflects the luminescence decay.

In the setup used for this method, the excitation pulse triggers the timing circuit (start signal) and the first photoelectron stops this circuit (stop signal). The circuit records the period between the start and stop signals and transforms it to the amplitude of the output signal (time–amplitude converter) The amplitude of signals is analyzed using a multichannel amplitude analyzer. As a result, kinetic curves are obtained whose ordinate corresponds to the number of pulses in the given channel of the analyzer, i.e., the luminescence intensity, and the abscissa corresponds to the channel number which is proportional to time.

As in other pulse methods, the resultant kinetic curves are convolutions of the true decay curves with the excitation pulse. Various methods have been proposed for deconvolution of these curves using computers and producing an elementary decay law upon δ-excitation. Owing to unavoidable noise, the deconvolution operation can generally only be carried out approximately. The problem is simplified if it is known in advance that the decay is exponential or can be approximated by a small number of exponents or by some predetermined function.

Fluorescence decays are determined in the Lebedev Physical Institute (Moscow) by time-correlated single photon counting (TCSPC), using a picosecond dye laser for excitation at 575 nm as described elsewhere [44]. The instrumental response function possesses a FWHM of 0.7 ns while the typical fluorescence lifetime for THIATS J-aggregates is 6 ns. All experiments are performed at very low excitation intensities (less then 10 photons/cm^2 per pulse) to avoid exciton–exciton annihilation and photodegradation of the samples.

The fluorescence lifetime can be obtained by analyzing decay curves within the framework of a multi-exponential model. The model tends to give values for χ^2 in the range of 1.05–1.50 for different temperatures. When the data are analyzed as the sum of two exponentials, χ^2 is usually below 1.30.

3.8 Application of Photon Counting to Kinetic Measurements on Nanoensembles

3.8.1 Dendrimer Rotation and Viscosity

Time-resolved fluorescence anisotropy measurements can provide important information about the rotation of nanosize ensembles (e.g., dendrimers) and assess the role of the environment. Dendrimers with probes are shown in Fig. 3.4. In fluorescence anisotropy decay experiments, the time-correlation function of the emission transition is determined. For details concerning the experiment and data analysis, see [72,73].

The excitation transition probability of a molecule which absorbs incident light is a function of the angle α between the directions of its absorption transition moment and the electric vector of the incident light, that is, $E(\alpha) = E_0 \cos^2 \alpha$. Therefore, fluorescence emitted from molecules with an opto-anisotropic field ought to be observed as remarkably polarized light. However, we cannot observe this polarized light in a general optical system.

The origin of the fluorescence depolarization phenomena can be classified into four categories: the gap between absorption and emission transition moments, intermolecular energy transfer, Brownian motion, and non-uniformity of fluorescent molecular orientation. Consequently, it is by the fluorescence depolarization method that we measure the degree of fluorescence depolarization and analyze particular physical factors based on the origin of the fluorescence depolarization phenomena in this experimental system. If there is no intermolecular energy transfer in an experimental system, we can extract only Brownian motion effects from the experimental emission anisotropy ratio data. In recent years, this technique has received attention as a way of measuring the Brownian motion of molecules and ensembles (or macromolecules) as this method has some useful properties, such as high sensitivity, molecular labeling, and real time measurement.

Assuming that the dendrimer rotates as a spherical particle, the overall rotational correlation time obtained from fluorescence anisotropy decay would

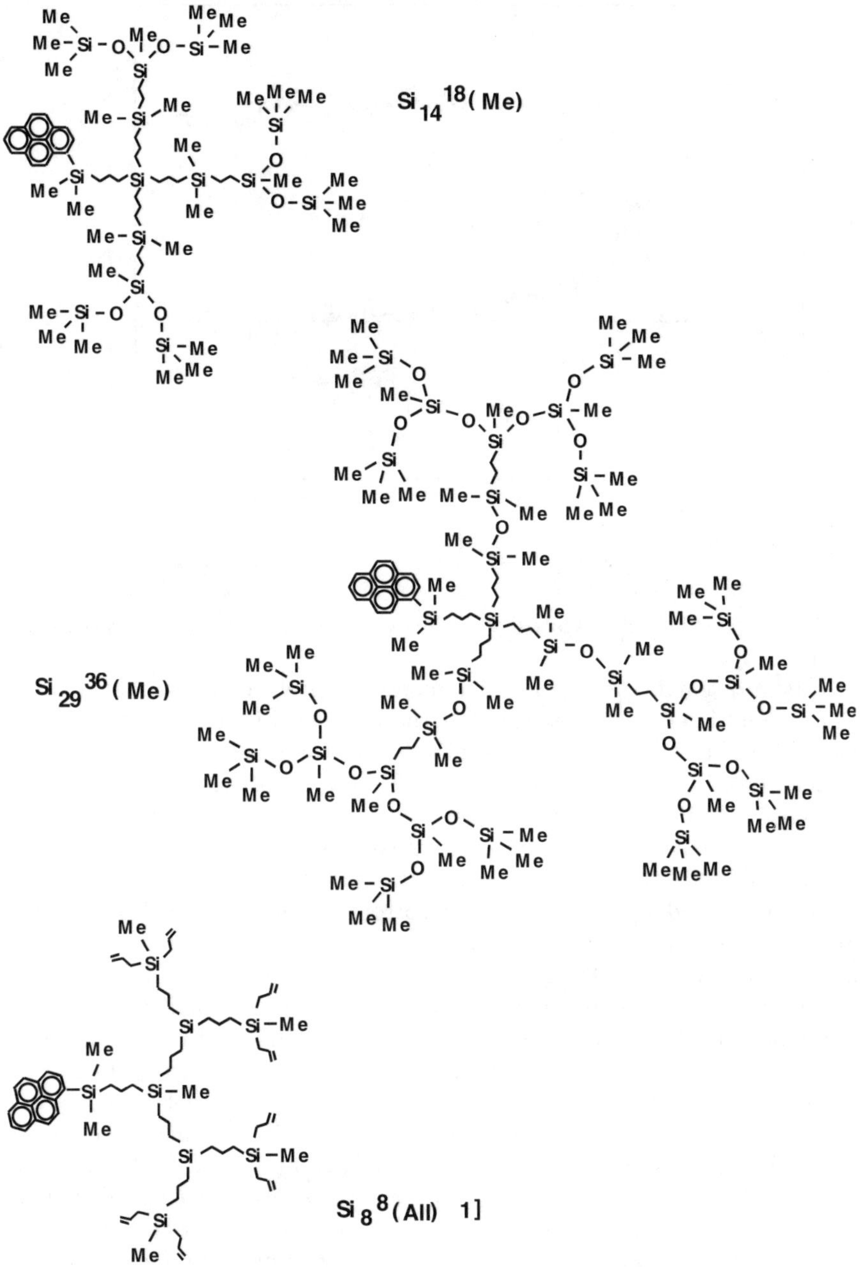

Fig. 3.4. Examples of a few generations of dendrimers: Si_{14}^{18}(Me), Si_{29}^{36}(Me), and Si_{8}^{8}(All)

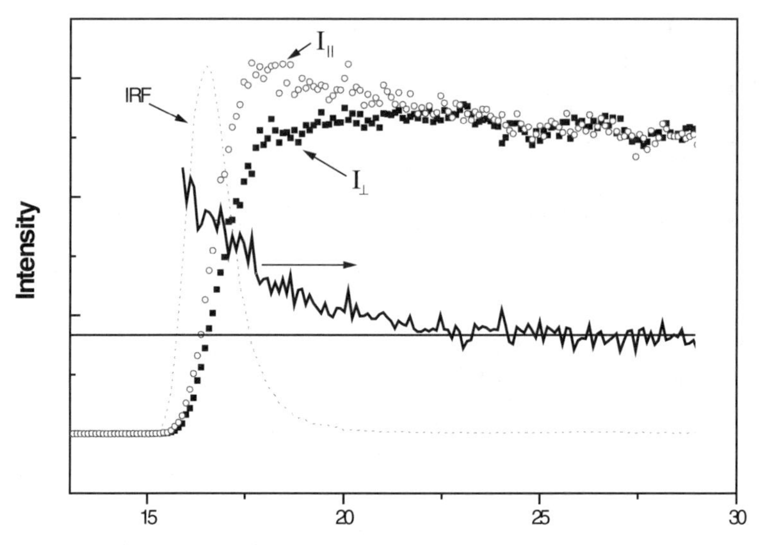

Fig. 3.5. Anisotropy of nanosecond kinetics

lead to the coefficient of viscosity of the environment via the Stokes–Einstein relationship.

The rotational reorientation of a fluorescent probe molecule that is bound to the dendrimer was investigated at the Lebedev Physical Institute with a view to evaluating the rotational mobility of the dendrimer. Polarized fluorescence decay was measured with an Edinburgh Instruments spectrofluorimeter Model 199 by the time-correlated single photon counting technique. Fluorescence steady-state spectra were recorded using a multichannel optical analyzer OMA-2 (EG&G) with a $\lambda = 337$ nm nitrogen laser.

We measured the fluorescence anisotropy decay of pyrenyl attached to Si_{14}^{18}(Me) and Si_{29}^{36}(Me) dissolved in cyclohexanol [74,75]. The kinetics of polarized fluorescence and fluorescence anisotropy of pyrenyl in Si_{29}^{36}(Me) is depicted in Fig. 3.5. The fluorescence anisotropy decays with a lifetime of about 2 ns reflecting the rotational reorientation of the fluorescent probe molecule attached to the dendrimer. It should be noted that in contrast to pyrenyl bound to Si_{29}^{36}(Me), the fluorescence for pure pyrenyl in cyclohexanol is completely depolarized and does not show any kinetics in the time range investigated. The fluorescence anisotropy decay for pyrenyl bound to Si_{14}^{18}(Me) was found to deviate from that for Si_{29}^{36}(Me) only at early times, coinciding at the tail of the decay curve. Additional experiments with better time resolution are required to elucidate the difference in reorientation mobilities of Si_{14}^{18}(Me) and Si_{29}^{36}(Me).

3.8.2 J-aggregate Radiative Lifetime and Structure

Experimental data which can give unbiased information about the dimension of J-aggregates are those concerning the temperature dependence of the radiative lifetime of the exciton [76,77]. To date, this relationship has been determined for only two systems: J-aggregates of 1,1′-diethyl-2,2′-cyanine (PIC) [78,79] and 1,1′-diethyl-3,3′-bis(sulfopropyl)-5,5,6,6′-tetrachlorobenzimidacarbocyanine (BIC) [80]. The temperature dependence of the fluorescence lifetime has also been measured for J-aggregates of some other dyes [81,82]. However, an accurate analysis of the data within the framework of the problem of exciton dimensionality has not been undertaken.

The effect of the temperature-induced population of exciton states has been calculated explicitly for a 1-D lattice by including the exciton–phonon interaction [83]. The transition dipole of a J-aggregate will also be limited by exciton scattering caused by static disorder. However, when these effects were considered in a theoretical analysis of the experimental data, it was impossible, within the framework of 1-D excitons, to explain the greater than 8-fold increase in τ_{rad} in PIC aggregates when the temperature is increased from 1.5 to 150 K [84]. An acceptable fit to the experimental data was obtained only within the framework of a 2-D exciton model, which allowed the authors to conclude that J-aggregates of PIC in solution possess a 2-D exciton coherent volume.

Results of photon counting investigations into the temperature dependence of the radiative lifetime of molecular aggregates are shown in Fig. 3.6. The main goal of the current photon counting application is to present and analyze new data for the temperature dependence of the radiative decay time obtained for J-aggregates of THIATS, which differ strongly from the well-known PIC aggregates.

To compare the temperature-dependent exciton dynamics of aggregates of THIATS to that of J-aggregates of PIC, it is convenient to consider the experimental data (radiative decay time) expressed in units of the inverse coherent length N_{c}^{-1}. The temperature dependence of the inverse coherent length observed for J-aggregates for THIATS and PIC is shown in Fig. 3.7. The main difference between these two systems is the rate of increase of N_{c}^{-1} with increasing temperature up to 130 K. For aggregates of THIATS, no strong temperature dependence of N_{c}^{-1} was observed.

To explain this slow increase of N_{c}^{-1} with temperature, it is not necessary to introduce assumptions about a possible 2-D character of the excitons; a 1-D crystal possesses enough ‘dark states’ which can become populated upon increasing the temperature. This example demonstrates the relation between dimensionality and experimental data obtained by the correlated single photon method.

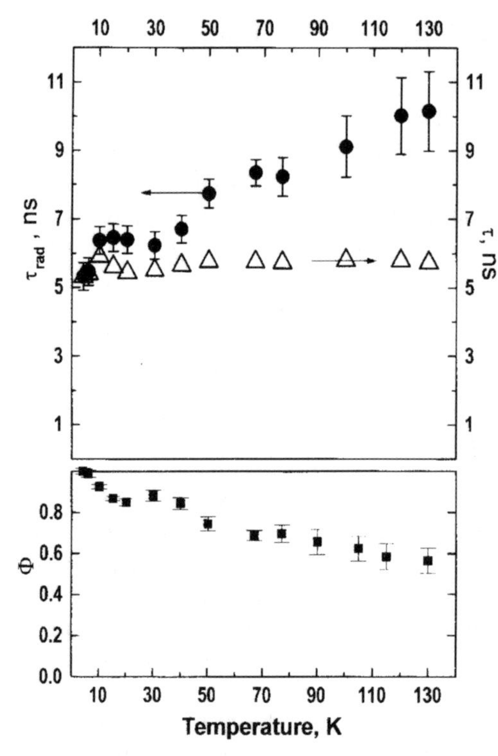

Fig. 3.6. *Top*: temperature dependence of the fluorescence decay time τ (*triangles* with size corresponding to experimental error) obtained upon excitation at 575 nm, and of the radiative exciton lifetime τ_{rad} (*circles*). *Bottom*: temperature dependence of the fluorescence quantum yield Φ for J-aggregates of THIATS in 3/2-water/ethylene glycol, with excitation at 532 nm

3.9 Photon Counting in SMS

By pulse laser excitation we can study fast electronic relaxation of excited nanostructures even if we deal with their ensembles. However, slow relaxation in the ground electronic state of nanostructures cannot be studied by pulse excitation of the ensembles. Slow relaxation on time scales from microseconds to days is typical for impurity centers in polymers and glasses [85,86]. Relaxation of the molecular ensemble in the ground electronic state will display some kind of averaged picture although experiments with single molecules reveal that the relaxation of each impurity center in an amorphous matrix displays individual features.

In order to carry out spectroscopy on single molecules, it is clear that we need to decrease the concentration of the guest molecule. When this concentration decreases, two circumstances work in favor of SMS. Firstly, distances between guest molecules increase. If the distance exceeds the size of the laser

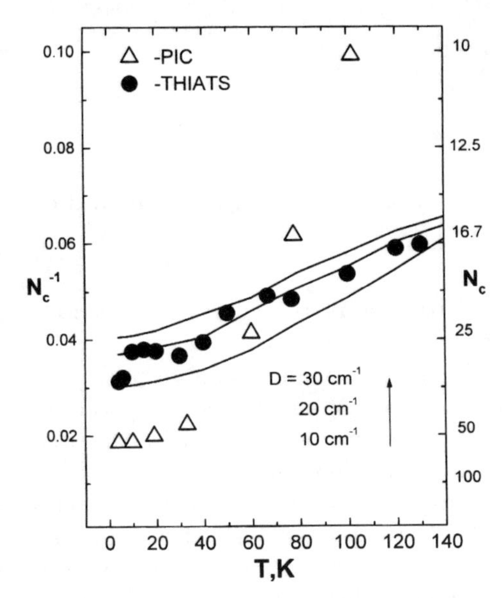

Fig. 3.7. Temperature dependence of inverse coherent length N_c^{-1} for J-aggregates of THIATS and PIC. *Solid lines* are theoretical curves for 1-D excitons

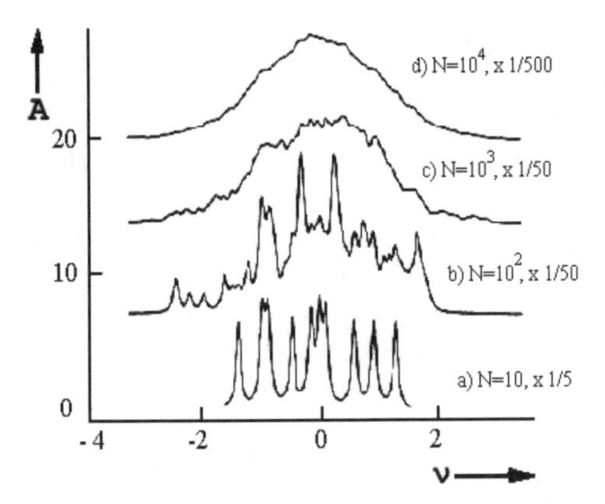

Fig. 3.8. Appearance of statistical fine structure in the inhomogeneously broadened optical band when the guest molecule concentration is reduced [87]

spot, the laser light will only excite a single molecule. Different types of near-field microscopy can reduce the size of the laser spot by a factor of ten compared with the diffraction limit. Secondly, when the guest molecule concentration is reduced, the inhomogeneously broadened optical band acquires a fine structure because in some frequency domains optical lines will be absent, as shown in Fig. 3.8.

It is obvious that large inhomogeneous broadening helps to reveal well resolved optical lines belonging to single molecules. Polymers and glasses

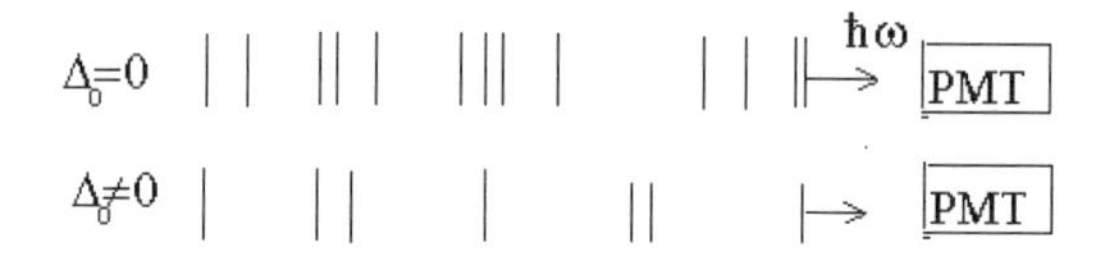

Fig. 3.9. Train of photons emitted by a single molecule under resonant and off-resonant laser excitation

serve as good solid matrices in SMS because of the large inhomogeneous broadening, reaching several hundred cm^{-1} in these matrices.

Since a single molecule can absorb only one photon, an absorption measurement is not feasible because it is difficult to detect the disappearance of one photon in excited laser light. It is possible only if special modulation methods developed for measuring weak absorption are used [5].

However, if the molecule studied has fluorescence, the simple method of excitation fluorescence can be used [6]. At the present time, this is the main method in SMS. In this case the molecule is embedded in a cooled polymer matrix and irradiated by CW monochromatic laser light. The molecule consequently absorbs and then emits one photon. A train of photons emitted by a single molecule is shown in Fig. 3.9.

In one-photon counting methods we count every photon from this train. The total number $N(\Delta_0)$ of photons counted by PMT for a predefined time interval, say one second, is proportional to the molecular absorption cross-section. It is a function of the detuning $\Delta_0 = \omega - \omega_0$, where ω_0 is the Bohr frequency of the molecular transition.

At short counting times the number $N(\Delta_0)$ is small and will fluctuate because absorption and emission events occur at random times. The ratio of the fluctuation $\delta N(\Delta_0)$ to the number $N(\Delta_0)$ of counted photons decreases as $\delta N(\Delta_0)/N(\Delta_0) = N(\Delta_0)^{-1/2}$ when $N(\Delta_0)$ increases. This means that the counting time is determined by the rate of photon emission and the desired signal-to-noise ratio.

By scanning the laser frequency, we can change the detuning Δ_0 and thereby measure the absorption line of a single molecule. An example of such an experiment is shown in Fig. 3.10.

This exemplifies the case when one-photon counting allows us to measure the probability of absorption of a photon with energy $\hbar\omega$. This measurement takes a long time if we hope to reach a good signal-to-noise ratio. Consequently, the impurity center must be stable. However, in polymers we deal with impurity centers of another type. These can change their resonant frequency as time goes by. Figure 3.11 shows just such an example. The line position was measured in a 2.5 s laser scan. Here we witness rare jumps between four spectral positions.

A random function $\omega(t)$ like the one shown in Fig. 3.11 is called a spectral trajectory. Spectral trajectories of various types have been observed, with fast

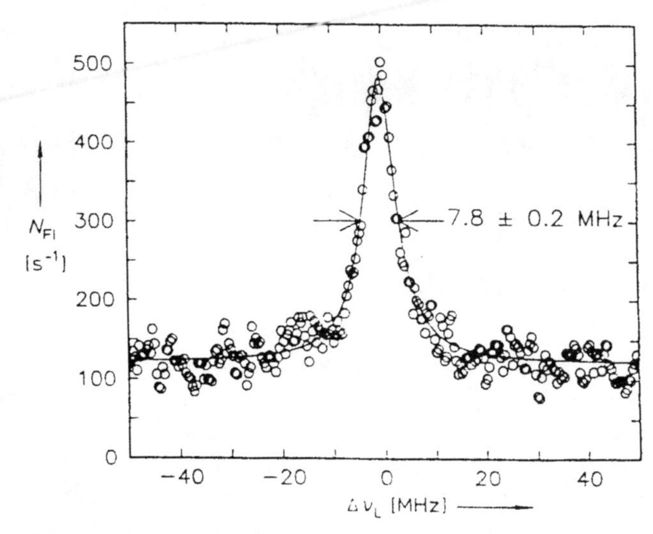

Fig. 3.10. Excitation fluorescence line of a single pentacene molecule in a para-terphenyl crystal at 1.5 K [88]

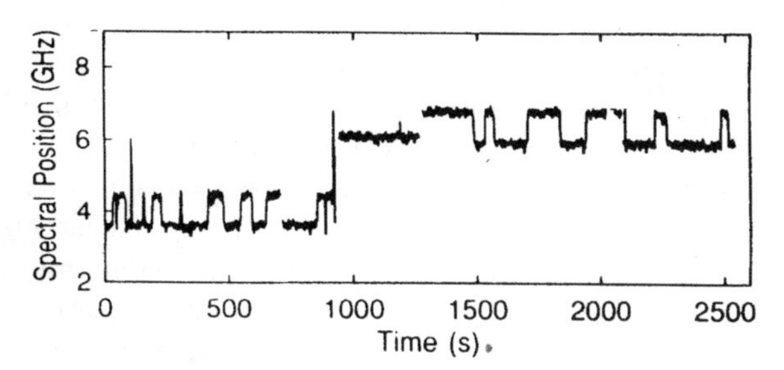

Fig. 3.11. Jumps in the absorption line of a single terrylene molecule doped polyethylene at 1.5 K [89]

and slow jumps between different spectral positions [8,89–93]. It is obvious that the spectral trajectory is a random function of time which contains information about some kind of relaxation processes. These processes are characterized by probabilities. Below we discuss the way these probabilities can be determined from the random trajectory.

3.10 Intermittency in Single-Molecule Fluorescence

The 'spectral trajectory' concept can be generalized by introducing the more general concept of a 'quantum trajectory'. Indeed any transition in a quantum system like a single molecule happens randomly in time. Therefore, any physical variable connected with the transition will fluctuate on the time

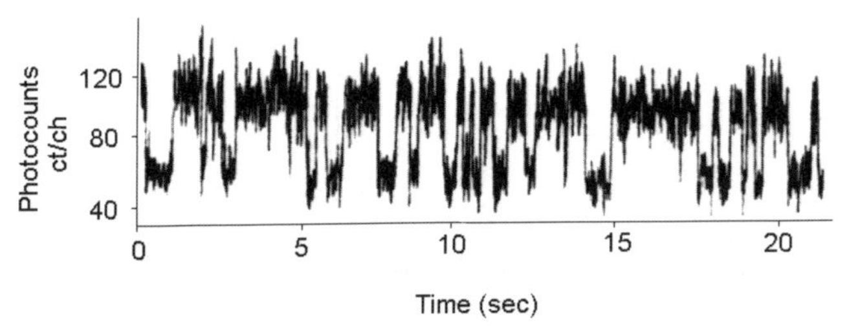

Fig. 3.12. Fluorescence from single FAD [92]

scale. The frequency fluctuations shown in Fig. 3.11 are only one of many possible quantum trajectory variations.

Figure 3.12 shows an example of quantum trajectory of another type when the trajectory describes fluorescence intensity fluctuations. These fluctuations resemble a random telegraph signal. This trajectory enables one to monitor the chemical reaction on a single-molecule level. This has been done by Lu, Xun and Xie in [92].

These authors studied enzymatic turnovers of oxidized single cholesterol molecules in real time by monitoring luminescence from the enzyme's fluorescence site, flavin adenine dinucleotide (FAD). FAD can emit light with a wavelength of 520 nm when in oxidized form. However, the fluorescence disappears after its reduction by cholesterol. Single FAD was irradiated by He–Cd laser light at a wavelength of 442 nm. Fluorescence was collected with an inverted fluorescence microscope by raster-scanning the sample. At large cholesterol concentrations, reduction and oxidation processes were in dynamic equilibrium. In the molecule ensemble, they observed equilibrium concentrations of both forms. However, on the single-molecule level, oxidation and reduction happen randomly in time, as can be detected via the appearance and disappearance of fluorescence. This random process is described by the quantum trajectory in Fig. 3.12. The lengths of 'off' and 'on' intervals take random values.

Lu et al. processed the measured quantum trajectories statistically, including up to 500 random off- and on-intervals. The histogram of on-intervals is shown in Fig. 3.13. The distribution of on-intervals can be described by the following function:

$$p(t) = \frac{ab}{b-a}\left[\exp(-at) - \exp(-bt)\right] , \tag{3.3}$$

where $a = 2.9\,\mathrm{s}^{-1}$ and $b = 17\,\mathrm{s}^{-1}$. This is a probability density for finding the molecule in oxidized form. It is characterized by two rate constants a and b which describe oxidation and reduction rates. This example shows how we can find the rate constants of the process from a quantum trajectory of random character.

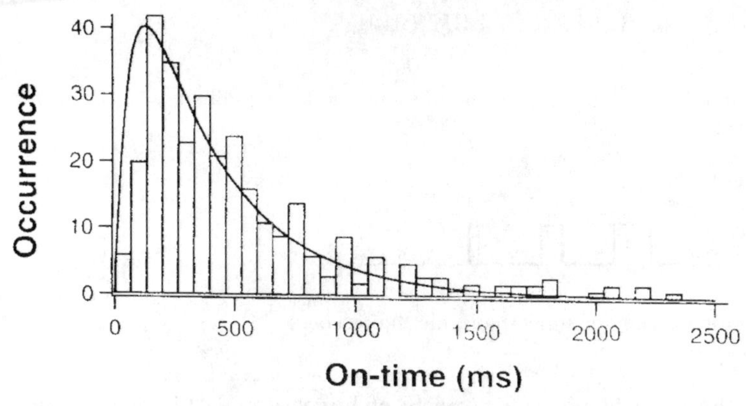

Fig. 3.13. The distribution of on-intervals in a quantum trajectory [92]

3.11 Spectroscopy of Single Molecule with Triplet Level. Photon Bunching

Figure 3.14 shows the energy diagram for the lowest electronic levels of a typical organic molecule. As a rule, these molecules have a triplet level between the ground state singlet level and the first excited singlet level. The probability γ_{ST} of transition between the triplet and the ground state singlet level is small so the triplet lifetime ranges from microseconds to seconds. The long-lived triplet level dramatically alters the train of emitted photons.

When the molecule irradiated by CW laser light lies in the singlet state, it can jump between the ground and excited singlet levels absorbing and emitting photons at random times. The probability Γ_{TS} of intersystem crossing in the excited electronic state is quite large and in fact is comparable with the probability $1/T_1$ of spontaneous photon emission. Therefore, the intersystem quantum yield $\eta = T_1\Gamma_{\mathrm{TS}}/(1 + T_1\Gamma_{\mathrm{TS}})$ is of the order of unity. When the excited molecule jumps to the triplet state 2, emission stops and radiation is absent while the molecule remains in the triplet state. After a random time of the order of $1/\gamma_{\mathrm{ST}}$, the molecule jumps to the ground electronic state and then emission appears again, as shown in Fig. 3.15.

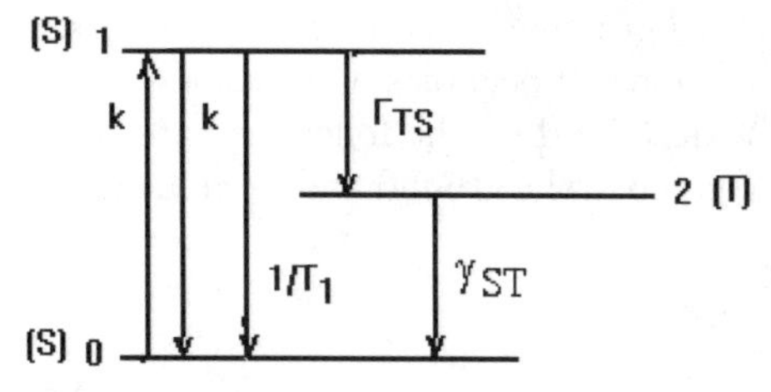

Fig. 3.14. Typical tree-level diagram for an organic molecule and rate constants which determine the dynamics of the molecule

II I III II IIII III IIII I I I II I III I I II II

Fig. 3.15. Photon bunching in photon emission due to the influence of the triplet electronic level

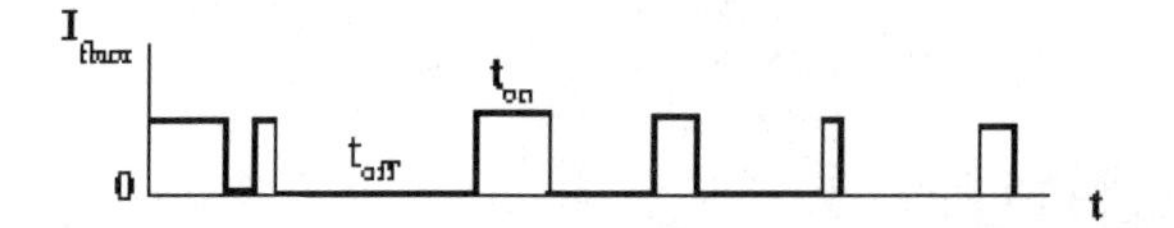

Fig. 3.16. Intermittency of single-molecule fluorescence

In fact, the train of photons consists of bunches and this phenomenon is called photon bunching. In accordance with Fig. 3.15, on-intervals when photons are emitted alternate with off-intervals t_{off} when there is no emission.

Let us now consider how to calculate probabilities for finding definite values of on- and off-intervals. Let the PMT count photons for a time interval t. If t is longer than the average interval between two photons, the PMT signal will be constant when fluorescence is emitted. When the molecule jumps to the triplet state, fluorescence disappears. We will measure a quantum trajectory, as shown in Fig. 3.16.

This quantum trajectory, which looks like a random telegraph signal, is related to the fluorescence intensity. From the trajectory, we can find average τ_{on} and τ_{off} intervals by substituting the measured t_{on} and t_{off} values for t_i in the formula

$$\tau = \frac{1}{N} \sum_{i=1}^{N} t_i \, . \tag{3.4}$$

A quantum trajectory can also characterize other physical variables like frequency, polarization, and so on. In all cases the quantum trajectory is a random function of time. Any trajectory is an irreproducible characteristic of the molecule and cannot only be calculated theoretically with the help of the Hamiltonian of the system. Indeed, quantum mechanics allows us to calculate probabilities but not trajectories. Let us consider how we can calculate the distributions of on- and off-intervals.

It is obvious when a molecule emits light that it is in singlet electronic states. When there is no emission the molecule occupies a triplet electronic level. The dynamics of singlet states is described by the following rate equations for probabilities of finding the molecule in the ground and excited singlet states:

$$\begin{aligned} \dot{\rho}_1 &= -\left(\frac{1}{T_1} + \Gamma_{\mathrm{TS}} + k\right)\rho_1 + k\rho_0 \, , \\ \dot{\rho}_0 &= \left(\frac{1}{T_1} + k\right)\rho_1 - k\rho_0 \, . \end{aligned} \tag{3.5}$$

Figure 3.14 explains the physical meaning of all the rate constants. It is obvious that

$$\rho_{\rm on} = \rho_1 + \rho_0 \tag{3.6}$$

is the probability of finding the molecule in a singlet electronic state, i.e., in an on-state. Summing the two equations, we find

$$\dot{\rho}_{\rm on} = -\Gamma_{\rm TS}\,\rho_1\,. \tag{3.7}$$

At times exceeding $1/\Gamma_{\rm TS}$, we may put $\dot{\rho}_1 = 0$ in (3.5). We then find from the first equation the relation

$$\rho_1 = \frac{kT_1}{1 + kT_1 + \Gamma_{\rm TS}T_1}\rho_0\,. \tag{3.8}$$

Making use of (3.7) and (3.8), we find

$$\rho_1 = \frac{kT_1}{1 + (2k + \Gamma_{\rm TS})T_1}\rho_{\rm on} \approx k\frac{T_1}{1 + \Gamma_{\rm TS}T_1}\rho_{\rm on}\,. \tag{3.9}$$

With the help of this equation, we can transform (3.7) to the form

$$\dot{\rho}_{\rm on} = -\frac{\rho_{\rm on}}{\tau_{\rm on}}\,, \tag{3.10}$$

where

$$\frac{1}{\tau_{\rm on}} = k\frac{\Gamma_{\rm TS}T_1}{1 + \Gamma_{\rm TS}T_1}\,. \tag{3.11}$$

The ratio here describes the intersystem crossing quantum yield. A solution of (3.14) is the function

$$w_{\rm on} = \frac{1}{\tau_{\rm on}}\exp\left(-\frac{t}{\tau_{\rm on}}\right)\,. \tag{3.12}$$

This is the probability density for finding a random value of $\mathrm{t_{on}}$ equal to t. It is a distribution function of on-intervals which can be measured after carrying out statistical processing of the quantum trajectory. It is obvious that $\tau_{\rm on}$ determined by (3.11) is an average value of on-intervals. It depends linearly on the laser light absorption intensity k.

When the molecule goes into the triplet state, emission stops. This means that the molecule is in the off-state. Hence we may write $\rho_2 = \rho_{\rm off}$. The dynamics of the off-state is described by the following equation:

$$\dot{\rho}_2 = -\gamma_{\rm ST}\rho_2\,, \tag{3.13}$$

$$\frac{1}{\tau_{\rm off}} = \gamma_{\rm ST}\,. \tag{3.14}$$

Solving (3.13), we find

$$w_{\rm off} = \gamma_{\rm ST}\exp(-\gamma_{\rm ST}t)\,. \tag{3.15}$$

This function describes a distribution of off-intervals. The value $\tau_{\rm off}$ determined by (3.14) is an average value of off-intervals and is independent of the intensity.

3.12 Two-Photon Correlators

By counting photons, we measure a quantum trajectory and then by processing the measured trajectory, we can find probabilities. A question arises: can we avoid the measurement of a random function like a quantum trajectory and measure transition probabilities directly by counting photons in real-time experiments? In fact, it is possible if we use the so-called two-photon counting method. As the name suggests, in these methods we count pairs of photons. The photon pair count rate is called the two-photon correlator. The theory for two-photon correlators has been developed recently in [93–95].

Figure 3.17 shows photons emitted by a single molecule irradiated by CW laser light. Let us consider pairs of photons with a definite time delay between them. The figure shows that there are two types of photon pair. For instance, we can count in the so-called start–stop regime by taking into account only those pairs consisting of consecutively emitted photons like pairs (2,3) and (12,13). The count rate $s(t)$ of such pairs is called the start–stop correlator. However, there is another possibility, namely, to count all pairs of photons with a delay, including additional pairs like the pair (4,6) with one intermediate photon 5. The count rate for all pairs with an arbitrary number of intermediate photons is called the full two-photon correlator. The count rate for photon pairs with a definite delay is proportional to the probability of finding this type of pair. It can be calculated quantum mechanically. The theory of two-photon correlators for two- and three-level molecules is developed in [93]. The theory has been generalized to molecules interacting with huge numbers of tunneling systems as happens in a polymer or glass matrix [94,95].

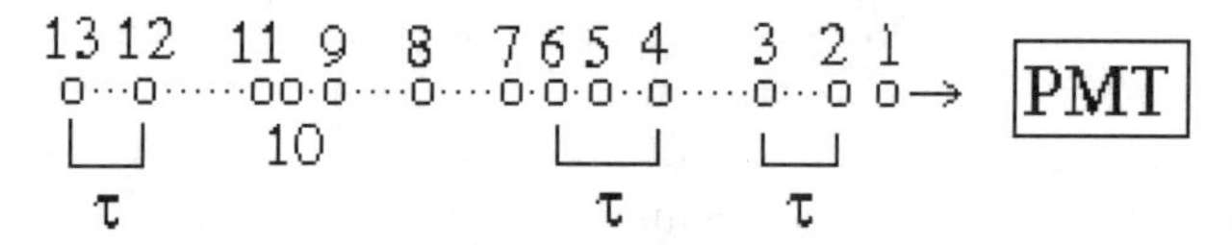

Fig. 3.17. Two types of photon pairs

Start–Stop Correlator. Let us begin by considering the simplest case relating to a two-level molecule interacting with a transverse electromagnetic field. Let W_0, W_1 and $W_{\boldsymbol{k}}$ be the probabilities of finding the molecule in the ground electronic state, in the excited electronic state and again in the ground electronic state after spontaneous emission of a photon with wave vector $\boldsymbol{k}$. If we take $t = 0$ as the time of recording the photon 2 shown in Fig. 3.17, we find that $W_0(0) = 1$ and $W_1(0) = W_{\boldsymbol{k}}(0) = 0$. Due to laser excitation, these probabilities will change. In time τ we find an emitted photon. It is obvious that

$$\mathrm{d}\sum_{\boldsymbol{k}} W_{\boldsymbol{k}}(\tau) = \sum_{\boldsymbol{k}} \dot{W}_{\boldsymbol{k}}(\tau)\,\mathrm{d}\tau \tag{3.16}$$

is the probability of photon 3 being emitted in the time interval from $t = 0$ to $t = \mathrm{d}\tau$. The count rate for pairs like the pair (2,3) is proportional to the probability $\sum_{\boldsymbol{k}} \dot{W}_{\boldsymbol{k}}(\tau)$, which reflects the molecular dynamics. It is shown in [93] that probabilities are governed by the conservation law

$$\dot{W}_0 + \dot{W}_1 + \sum_{\boldsymbol{k}} \dot{W}_{\boldsymbol{k}} = 1\,. \tag{3.17}$$

This means that one of the probabilities need not be considered. It was also shown that

$$\frac{W_1(t)}{T_1} = \sum_{\boldsymbol{k}} \dot{W}_{\boldsymbol{k}}(t) = s(t)\,, \tag{3.18}$$

and that the probability W_1 can be found from the following equations

$$\begin{aligned}
\dot{W}_{10} &= -\mathrm{i}(\Delta - \mathrm{i}\Gamma)W_{10} - \chi(W_0 - W_1)\,, \\
\dot{W}_{01} &= \mathrm{i}(\Delta + \mathrm{i}\Gamma)W_{01} - \chi(W_0 - W_1)\,, \\
\dot{W}_1 &= -\chi(W_{10} + W_{01}) - \frac{W_1}{T_1}\,, \\
\dot{W}_0 &= \chi(W_{10} + W_{01})\,,
\end{aligned} \tag{3.19}$$

where Δ is the difference between the laser and resonant frequencies, $\Gamma = 1/2T_1$, and $\chi = \boldsymbol{E} \cdot \boldsymbol{d}/h$ is the Rabi frequency. Equation (3.18) can easily be derived by summing the third and fourth equations in (3.19) provided we take into account (3.17). The function $s(t)$ is called the start–stop correlator.

Full Two-Photon Correlator. If we count all pairs with a time delay and take into account the pair (4,6) with one intermediate photon (Fig. 3.17) and also other pairs with various numbers of intermediate photons as shown in Fig. 3.18, we arrive at the full two-photon correlator. The first term in Fig. 3.18 relates to the start–stop regime where there are no intermediate

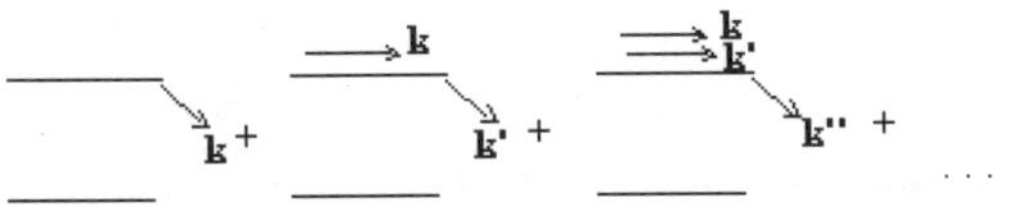

Fig. 3.18. Summation of photon pairs with various numbers of intermediate photons

photons. The second term relates to the pair with one intermediate photon $\boldsymbol{k}$, and so on. The counting rate in this case equals [93]

$$\begin{aligned}\sum_{\boldsymbol{k}} \dot{W}_{\boldsymbol{k}} + \sum_{\boldsymbol{k}\boldsymbol{k}'} \dot{W}_{\boldsymbol{k}\boldsymbol{k}'} + \sum_{\boldsymbol{k}\boldsymbol{k}'\boldsymbol{k}''} \dot{W}_{\boldsymbol{k}\boldsymbol{k}'\boldsymbol{k}''} + \ldots &= \frac{1}{T_1}\left(W_1 + \sum_{\boldsymbol{k}} W_{1\boldsymbol{k}} + \ldots\right) \\ &= p(t) = \frac{\rho_{11}}{T_1}\,, \end{aligned} \tag{3.20}$$

where ρ_{11} satisfies the optical Bloch equations

$$\begin{aligned} \dot{\rho}_{10} &= -\mathrm{i}(\Delta - \mathrm{i}\Gamma)\rho_{10} - \chi(\rho_{00} - \rho_{11})\,, \\ \dot{\rho}_{01} &= \mathrm{i}(\Delta + \mathrm{i}\Gamma)\rho_{10} - \chi(\rho_{00} - \rho_{11})\,, \\ \dot{\rho}_{11} &= -\chi(\rho_{10} + \rho_{01}) - \frac{\rho_{11}}{T_1}\,, \\ \dot{\rho}_{00} &= \chi(\rho_{10} + \rho_{01}) + \frac{\rho_{11}}{T_1}\,. \end{aligned} \tag{3.21}$$

Equation (3.20) determines the full two-photon correlator.

Using (3.19) and (3.21), we can a find simple relation between the start–stop and full two-photon correlators [93]

$$p(t) = s(t) + \int_0^t s(t-x)p(x)\,\mathrm{d}x\,. \tag{3.22}$$

Solving this integral equation by an iteration procedure, we find the solution in the form of a sum with an infinite number of terms. These terms relate to the terms of the sum in (3.20). For instance, the first term

$$\sum_{\boldsymbol{k}} \dot{W}_{\boldsymbol{k}} = s(t) \tag{3.23}$$

describes the counting rate in the start–stop regime, whilst the second term

$$\sum_{\boldsymbol{k}\boldsymbol{k}'} \dot{W}_{\boldsymbol{k}\boldsymbol{k}'} = \int_0^t s(t-x)s(x)\,\mathrm{d}x \tag{3.24}$$

describes the contribution to the full two-photon correlator from photon pairs like the pair (4,6) with one intermediate photon 5, as shown in Fig. 3.17. It is obvious that the nth term of the sum describes the contribution from pairs with n intermediate photons.

Figure 3.19 shows start–stop and full correlators calculated in [93] with the help of (3.19) and (3.21) for various values of the Rabi frequency. The start–stop correlator aproaches zero on time scales similar to the fluorescence time T_1. It is impossible to research the slow dynamics typical of chromophores in polymers and glasses by means of the start–stop correlator. Slow dynamics in polymers and glasses must therefore be probed by the full two-photon correlator.

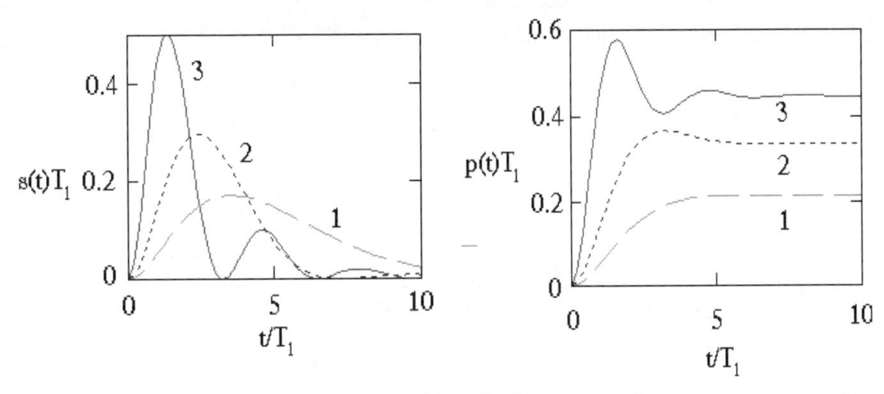

Fig. 3.19. Dependence of correlators $s(t)$ and $p(t)$ on time delay between two photons at $\Delta_0 = 0$ and $\chi T_1 = 0.1$ (1), 0.2 (2) and 1 (3)

Because we cannot find photon pairs with very short delays, both types of correlators approach zero at these short delays. This phenomenon is called anti-bunching. The full two-photon correlator is a constant at large times. This means the absence of any correlation between two photons with large delay. The second photon 'forgets' about the first photon if the time delay is much longer than the fluorescence time T_1. However, if some kind of slow process exists in the system and this process influences absorption, the probability of finding the second photon will feel this slow process. It will be reflected in the long-term behavior of the full correlator. In the next section, we consider just such an example.

3.13 Full Two-Photon Correlator for Single Molecule with Triplet Level

We have already seen in Sect. 3.11 that the triplet level strongly influences the photon train emitted by a single molecule irradiated by CW laser light. Let us calculate the full two-photon correlator for such a single molecule. In order to describe this case, we need to add a fifth equation to the four we already have in (3.21). This equation should describe a transition from the excited singlet level to the triplet level and a transition from the triplet level to the ground electronic state in accordance with the diagram in Fig. 3.14. The rate of light-induced transitions is

$$k(\Delta_0) = 2\chi^2 \frac{1/T_2}{\Delta_0^2 + 1/T_2^2} , \tag{3.25}$$

where Δ_0 is the difference between the laser frequency and the Bohr frequency of the singlet–singlet resonant transition. The expression for the full two-photon correlator for such a molecule was found in [93]:

$$p(t) = \frac{k}{T_1}\left[\frac{\gamma_{\mathrm{ST}}}{\gamma_0^2 - R_0^2} + \left(1 - \frac{\gamma_{\mathrm{ST}}}{\gamma_0 - R_0}\right)\frac{\mathrm{e}^{-(\gamma_0 - R_0)t}}{2R_0} - \left(1 - \frac{\gamma_{\mathrm{ST}}}{\gamma_0 + R_0}\right)\frac{\mathrm{e}^{-(\gamma_0 + R_0)t}}{2R_0}\right], \quad (3.26)$$

where

$$\gamma_0 = \frac{\Gamma + 2k + \gamma_{\mathrm{ST}}}{2}, \quad R_0 = \sqrt{\left(\frac{\Gamma + 2k - \gamma_{\mathrm{ST}}}{2}\right)^2 - \Gamma_{\mathrm{TS}} k}\,. \quad (3.27)$$

The temporal dependence of this correlator is shown on a logarithmic time scale in Fig. 3.20.

Specific features of this correlator have a definite physical meaning. The value of the ordinate at small delays, say 10^{-5} s, corresponds to the count rate of photon pairs in a bunch. The value of the ordinate at long delays, say 1 s, corresponds to a situation where we count photon pairs consisting of photons from various branches. In other words, it is the count rate for photon branches.

Exponential decay looks like a smooth step over one order of magnitude of time on the logarithmic time scale. The two steps on the curve in Fig. 3.20 can be related to the exponential growth of emission with time delay 10^{-7}s and exponential decrease in emission with time delay 10^{-2}s. These two times correspond to the rate of population of excited electronic level 1 and to its rate of depopulation due to the transition to triplet level 2 in Fig. 3.14. This example clearly shows that the full two-photon correlator enables us to investigate slow relaxation in the system. This is a very important feature of the full two-photon correlator because it can be used to investigate slow relaxations in polymers and glasses.

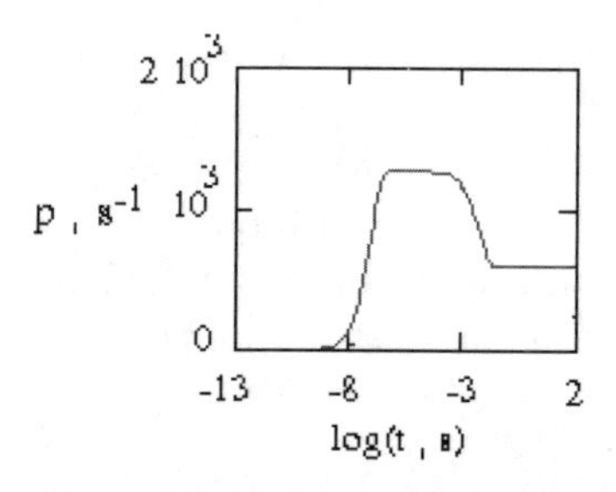

Fig. 3.20. Full two-photon correlator for a single molecule with triplet level calculated by means of (3.26) with $T_2/T_1 = 10^{-2}$, $\gamma_{\mathrm{ST}}T_2 = 10^{-6}$, $\Gamma_{\mathrm{TS}}T_1 = 9$ and $\chi T_2 = 2 \times 10^{-4}$

3.14 Tunneling Dynamics of Polymers Probed by SMS

It is well known that low temperature properties of polymers are due to the existence of so-called two-level systems (TLS) [96,97]. They are described by a double-well potential as shown in Fig. 3.21.

The change in the energy splitting of the double-well adiabatic potential resulting from electronic excitation of the chromophore is the parameter of a chromophore–TLS interaction of the Franck–Condon type. This chromophore–TLS interaction determines anomalous temperature broadening of spectral holes burned in the chromophore optical bands. Such broadening depends linearly on temperature in the mK temperature range [98] in contrast to the temperature broadening in crystals which follows either T^7 or Arrhenius-type temperature laws [99].

In accordance with Fig. 3.21, the optical band of the chromophore interacting with a TLS consists of two optical lines with separation Δ in the frequency scale.

These doublets could serve as direct evidence for the existence of TLSs in polymers. Nevertheless, they have never been observed in hole-burning spectroscopy of polymers because there is a distribution over frequency separations in molecular ensembles. The first direct observation of such doublets was made in SMS [100].

There are seas of TLSs in every polymer. Their relaxation rates R are distributed over a very wide time scale, covering more than ten orders of magnitude from microseconds to days. It is clear that the slow relaxation of TLSs must manifest itself in the full two-photon correlator. The theory for two-photon correlators [93] was therefore generalized to take into account the interaction of a single chromopore with a sea of TLSs. This was done in [94,95].

The theory was built up by taking into account the following hierarchy of rates

$$1/T_1 \gg k \gg A\,,\ a\,,\ B\,,\ b\,, \tag{3.28}$$

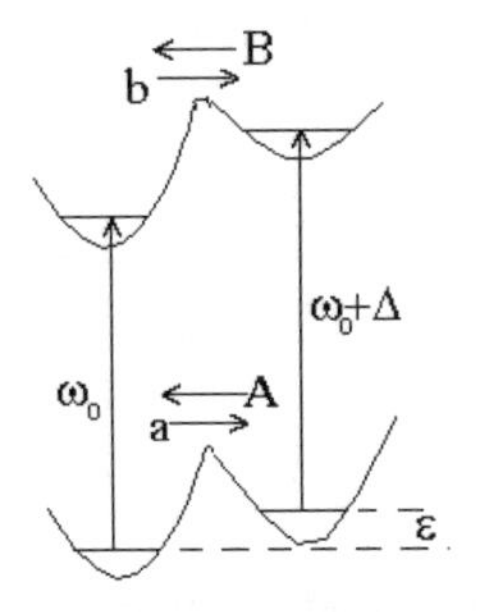

Fig. 3.21. Tunneling system of a polymer in the ground and excited electronic states of a chromophore

as shown in Fig. 3.21. Here T_1 is the chromophore lifetime, k is the rate of light-induced transitions and A, a, B, and b are rates of tunneling in the ground and excited electronic states of a chromophore. As shown in [94,95] tunneling rates in the excited electronic state of the chromophore are proportional to kBT_1 and kbT_1. Hence if A, a, B and b are of the same order, we can neglect light-induced tunneling in the excited electronic state. This conclusion is correct for all the TLSs inherent in a polymer. In this case the formula for the full two-photon correlator takes the form

$$p(t) = k(\Delta_0, t, T)\left(1 - \mathrm{e}^{-t/T_1}\right), \tag{3.29}$$

where the function $k(\Delta_0, t, T)$ of detuning $\Delta_0 = \omega - \omega_0$ and temperature T describes the light absorption coefficient of the chromophore interacting with TLSs, some of which had not reached thermal equilibrium by the time t when the second photon of the photon pair was emitted. For the case when a chromophore interacts with a single TLS, this function takes the form

$$k(\Delta_0, t, T) = \left[1 - \rho(t, T)\right]k(\Delta_0) + \rho(t, T)k(\Delta_0 - \Delta), \tag{3.30}$$

where $k(\Delta_0)$ is a Lorentzian described by (3.25), and

$$\rho(t, T) = f(T) + [\rho(0) - f(T)]\exp(-Rt) \tag{3.31}$$

determines the probability of finding the TLS in the excited state with the energy ε. Here $R = A + a$. It is obvious that the probability in thermal equilibrium is $\rho(\infty, T) = f(T) = [\exp(\varepsilon/kT) + 1]^{-1}$.

How will TLS relaxation manifests itself in the two-photon correlator? The full two-photon correlator calculated using (3.29) and (3.30) is shown in Fig. 3.22. For excitation at frequency ω_0 we must take $\rho(0) = 0$ in (3.31) because at this moment of time the system occurs in the left-hand well of the double-well potential shown in Fig. 3.21. For excitation at frequency $\omega_0 + \Delta$, we must take $\rho(0) = 1$ in (3.31).

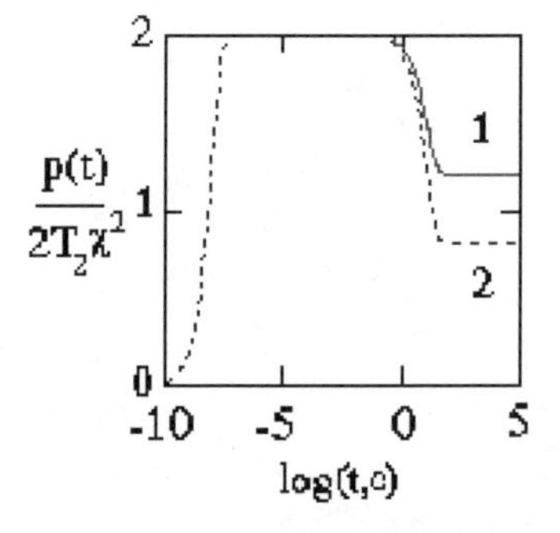

Fig. 3.22. Temporal dependence of full two-photon correlators and excited electronic state of a guest molecule for excitation at frequencies ω_0 (1) and $\omega_0 + \Delta$ (2). $f = 0.4$, $\Delta = 8/T_2$, $T_1 = 10^{-8}$ s and $R = 10^{-1}\mathrm{s}^{-1}$

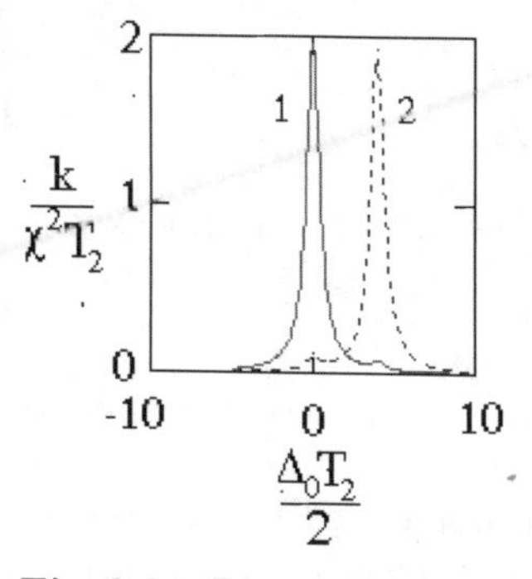

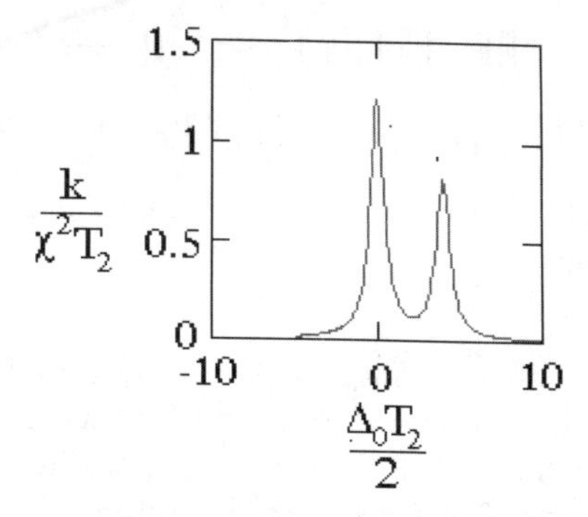

Fig. 3.23. Dependence of two-photon correlator on detuning for the same values of parameters as in Fig. 3.22. *Left*: at short delays with $tR \ll 1$. *Right*: at long delays with $tR \gg 1$

The correlators shown in Fig. 3.22 increase their values at times of order T_1 and decrease their values at times of order $1/R$ which characterizes TLS relaxation. The ratio of the correlators 1 and 2 at long delays is approximately equal to the ratio $(1 - f)/f$. Comparison of Figs. 3.22 and 3.20 reveals that relaxation due to singlet–triplet transitions and relaxation due to tunneling transitions manifest themselves in the full two-photon correlator in a similar fashion. Let us consider now the correlator described by (3.29) and (3.30) as a function of the laser frequency.

The result of our calculation is shown in Fig. 3.23. These pictures of the two-photon correlator with short and long delays between photons are convenient when discussing spectral pictures obtained in one-photon experiments with short and long laser scans.

For fast laser scans, when $t_{sc}R \ll 1$, we observe a single optical line jumping randomly between two spectral positions from one laser scan to another. The spectral picture is similar to that shown in Fig. 3.23 (left). Which line will appear is determined by a random factor, depending on whether the TLS is found in the left- or right-hand well.

For slow laser scans, when $t_{sc}R \gg 1$, we observe just two optical lines in every laser scan as shown in Fig. 3.23 (right). The line intensities are proportional to the probabilities $1 - f$ and f of finding the TLS in the left- or right-hand well. For slow scans, we cannot detect relaxation if it is faster than the duration of the scan. The duration of the laser scan determines the time resolution of the experimental setup. In STM measurements of the I–V characteristic the time of the current measurement plays the role of t_{sc} in optical experiments. However, it is worth noting that in two-photon experiments the shortest delay t between two photons will characterize the time resolution of the setup. This delay can be several orders of magnitude shorter than the best time resolution in experiments with electric current.

The physical picture discussed in connection with Fig. 3.23 will manifest itself in the train of emitted photons. This train of photons is shown in

ω_0

$\omega_0+\Delta$

Fig. 3.24. Train of photons emitted by a single molecule excited with various laser frequencies

Fig. 3.24. When a single molecule is excited at frequency ω_0, a pause appears in the emission at the random time when the TLS jumps from the left- to the right-hand well. The chromophore in the right-hand well has a different resonant frequency, $\omega_0 + \Delta$. Therefore, if the laser frequency is not tuned to the new value of the resonant frequency, the emission disappears and the lower train of photons in Fig. 3.24 will be absent. In this special case Fig. 3.24 will be similar to Fig. 3.15 which shows the train of photons emitted by a single molecule with a triplet level. If we tune the laser frequency to $\omega_0 + \Delta$, we will observe bunches from the lower train in Fig. 3.24. It is important to note that pauses in one train of photons exactly coincide with bunches in the other train, and vice versa. This is a telling feature of the chromophore interacting with a TLS.

3.15 Relation Between One- and Two-Photon Counting Methods in a More Complex Case

In the preceding section we considered the relationship between one- and two-photon methods for the simplest case when a chromophore interacts with a single TLS. The relevant spectral trajectory will display random jumping between two spectral positions. It is obvious that this physical model cannot describe the spectral trajectory shown in Fig. 3.11 where the optical line jumps between four spectral positions. This trajectory can be understood with the help of a model which describes a chromophore interacting with two TLSs.

Two TLSs have four levels. Therefore, instead of (3.30), we should take a coefficient k with the form

$$\begin{aligned} k(\Delta_0, t, T) = \big[&(1-\rho)(1-\rho')k(\Delta_0) + \rho(1-\rho')k(\Delta_0 - \Delta) \\ &+ (1-\rho)\rho' k(\Delta_0 - \Delta') + \rho\rho' k(\Delta_0 - \Delta - \Delta')\big] , \end{aligned} \tag{3.32}$$

where ρ and ρ' describe the probabilities of finding the first and second TLS to be excited. They are described by equations similar to (3.30). The Lorentzian $k(\Delta_0)$ is described by (3.25). The frequency of the laser light determines initial conditions for the probabilities ρ and ρ'. For example, if we excite a single molecule with the detuning $\Delta_0 = \omega - \omega_0 = 0$, only the first Lorentzian in (3.32) works. Consequently, the system consisting of a chromophore with two TLSs occurs after emission of a photon in the quantum state with $\rho(0) =$

$\rho'(0) = 0$. If we excite the system via the second Lorentzian in (3.32) we must take $\rho(0) = 1$, $\rho'(0) = 0$, and so on.

Figure 3.25 shows the results of calculating the full two-photon correlator by means of (3.32) upon excitation via the first, second, third and fourth Lorentzian, i.e., for the initial conditions $\rho(0) = \rho'(0) = 0$, $\rho(0) = 1$ and $\rho'(0) = 0$, $\rho(0) = 0$ and $\rho'(0) = 1$, and $\rho(0) = \rho'(0) = 1$. The curves in Fig. 3.25 were calculated for three different time delays between photons.

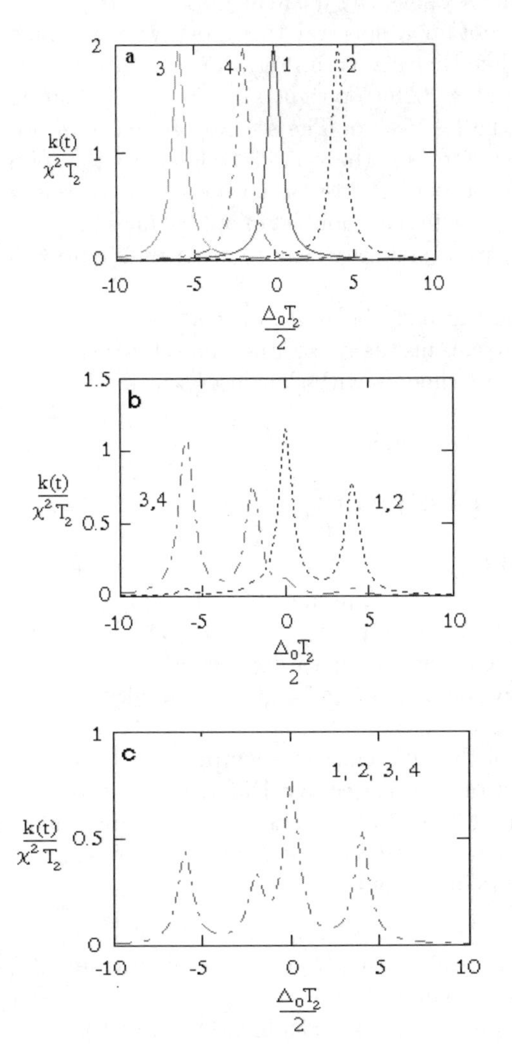

Fig. 3.25. Full two-photon correlator upon excitation via various optical lines and at various time delays between photons $t = 10^{-5}$ s (*top*), $t = 1$ s (*middle*), $t = 10^2$ s (*bottom*). TLS parameters are as follows: $\varepsilon = 0.5$ K, $R = 6 \times 10^2 \mathrm{s}^{-1}$, $T_2\Delta = 4$, $\varepsilon' = 1$ K, $R' = 10^{-1} \mathrm{s}^{-1}$, $T_2\Delta' = -6$, $T = 1.7$ K

For $t < 1/R < 1/R'$ we obtain Fig. 3.25 (top left). The relaxation in both TLSs is very slow and relaxation takes longer than the delay between photons in the pair. In this case the correlator as a function of the frequency is described by a single Lorentzian. Which of the four Lorentzians will manifest itself depends on the value of the laser frequency. As already shown in the preceding section, the spectral picture for the two-photon correlator can be related to a spectral trajectory measured in one-photon experiments with laser scans. If the laser scan time satisfies the inequality $t_{\mathrm{sc}} < 1/R < 1/R'$ in one-photon experiments, we obtain a spectral trajectory with jumping between four spectral positions like that shown in Fig. 3.11.

For $1/R < t < 1/R'$ we obtain Fig. 3.25 (top right). In this case relaxation in one TLS occurs faster than the laser scan. The second photon does not feel this fast relaxation. We therefore see the doublet related to the TLS with fast relaxation upon excitation via 1,2 lines or 3,4 lines. In one-photon experiments, if $1/R < t_{\mathrm{sc}} < 1/R'$ we will see a doublet of optical lines jumping randomly on the frequency scale from one laser scan to another. Figure 3.25 (top right) shows this jumping.

For $1/R < 1/R' < t$ we obtain Fig. 3.25 (bottom). In this case, we will see all four optical lines in each laser scan just as we see the same stable spectral picture in one-photon counting experiments with slow laser scans.

3.16 Local Dynamics of Individual Molecules in Polymers

The spectral trajectory shown in Fig. 3.11 is just one example from the wide variety of spectral trajectories measured in experiments with single terrylene molecules in polyethylene crystals. In fact every guest molecule has its own individual spectral trajectory even when used to dope crystals [88,89]. It reveals the wide range of local environments even for guest molecules existing in crystals.

It is worth noting that some of the differences in spectral trajectories can result from the random character of each trajectory. Differences of this type do not reflect any differences in the local environment of guest molecules. Only differences in probabilities measured in experiments can prove that there really is a variety of local environments.

Line Width Distribution. The half-width $2/T_2$ of a zero-phonon line depends on temperature and exceeds the fluorescence rate $1/T_1$. This half-width results from electron–phonon and electron–TLS coupling. One might not expect such variety in this type of coupling for guest molecules in crystals. However, SMS revealed that a distribution over half-widths exists even in the anthracene single crystal [101]. Such distributions over line widths of single molecules are larger in polymers. An example is shown in Fig. 3.26.

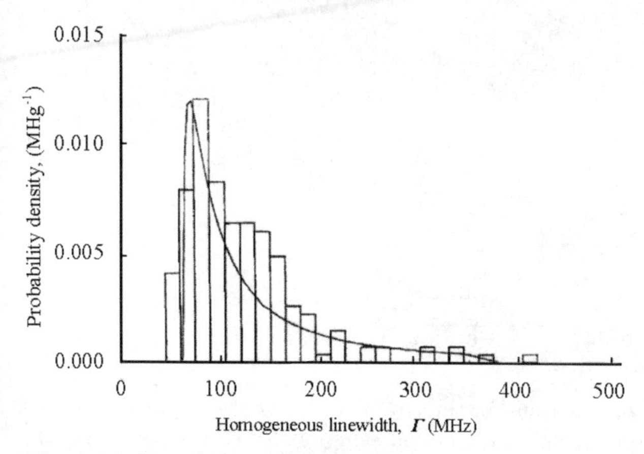

Fig. 3.26. Distribution of the 176 line widths of terylene molecules in a polyethylene matrix at $T = 1.8\,\mathrm{K}$ [90]

For terylene molecules doping the Shpol'ski matrix, poly (vinyl butyral), poly (methyl ethacylate) and polystyrene, such distributions can be found in [101]. The existence of distributions over line widths means that optical dephasing of single molecules is determined by the interaction of the guest molecule with localized phonons and TLSs in the immediate neighbourhood of the molecule.

Temporal Behavior of Full Two-Photon Correlators. The temporal behavior of the two-photon correlator characterizes probabilities of relaxation processes. Therefore differences in two-photon correlators can be related to differences in the local environment of individual guest molecules. In experiments with single molecules, an autocorrelation function (AF) $g^{(2)}(t)$ is measured. The AF is related to the fluorescence intensity $I(t)$ and can be expressed via the full two-photon correlator as follows:

$$\frac{p(t_0)}{p(\infty)} = g^{(2)}(t_0) = \lim_{t\to\infty} \frac{\langle I(t)I(t+t_0)\rangle}{\langle I(t)I(t+\infty)\rangle} = \lim_{t\to\infty} \frac{\langle I(t)I(t+t_0)\rangle}{\langle I(t)\rangle^2} \,. \tag{3.33}$$

The infinite time is determined in practice as the time when all relaxation is over. The AF thus approaches unity at large time delays t_0.

In practice, it is more convenient to use the AF than the two-photon correlators, because all factors relating to experimental conditions beyond the control of the researcher are excluded by taking a ratio. However, if the count rate of photon pairs with large delay is very small, the value of $p(\infty)$ is very small as well. Therefore, the value of the AF will be found with large error. This type of situation emerges when we count photons emitted by a single molecule with triplet level. If the probability γ_{ST} of the triplet–singlet transition is small, the value of $p(\infty)$ will be small and it will be measured

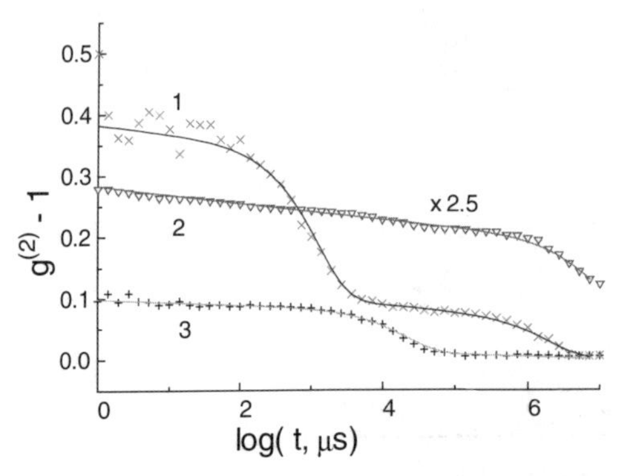

Fig. 3.27. Autocorrelation functions for three terylene molecules in polyethylene at $T = 1.8\,\mathrm{K}$ [90] and theoretical curves calculated using (3.29), (3.30) and (3.32) [31]

with large inaccuracy. This will dramatically influence the ratio in (3.33). The situation can become worse still if a molecule has undergone spectral-hole burning. In this case emission disappears at large times. It is then unclear what value of $p(\infty)$ we should use in (3.32).

Figure 3.27 shows the AF of three individual terylene molecules in a polyethylene matrix. They are quite different. Since the formula for the AF involves probabilities, the difference can be related to a difference in the immediate neighbourhoods of the three guest molecules. The curves show the difference in local dynamics for three individual molecules.

On a logarithmic time scale, a smooth step by one order of magnitude means an exponential decay. Therefore, we may say that in the immediate vicinity of molecule 1, there are two TLSs with relaxation rates $10^3\,\mathrm{s}^{-1}$ and $1\,\mathrm{s}^{-1}$, whilst there is one TLS with relaxation rate $0.1\,\mathrm{s}^{-1}$ and $100\,\mathrm{s}^{-1}$ in the vicinity of molecules 2 and 3, respectively.

3.17 Conclusions

3.17.1 Fast Photon Counting Measurements vs. CW Measurements

The time dependence of light emission from photoexcited molecules is studied extensively in physics and chemistry to understand the interactions and reactions of those molecules. However, until now, nearly all such experiments have studied large numbers of molecules (ensembles) simultaneously, thus providing data that represents the average of the behavior of the individual molecules. This severely limits the study of these phenomena.

The new approach involves a matrix containing fluorescent molecules and exposed to focused laser light. The molecules are spatially separated in such a way that only one is exposed to the light at any time. When the sample is scanned with the laser, a fluorescence image can be obtained in which each bright spot corresponds to a single molecule.

By continuously sweeping the laser polarization from 0 to 90 degrees while examining a single molecule, the orientation of the dipole of the molecule was determined with high precision and its changes in time recorded.

In some cases, the light emission of a single molecule is observed to switch on and off as the molecule is presumed to convert between light-emitting and non-light-emitting or 'dark' states. By measuring the dipole orientation before and after the 'dark state jump', researchers have been able to show that the molecule remains stationary as it switches state. In other cases researchers observe changes in the polarized emission signal indicating that the molecules experienced 'rotational jumps' in which they reoriented themselves.

The fundamental difference between optical studies of molecule ensembles (molecular aggregates or macromolecular dendrimers) and optical studies of single molecules should be emphasized. In the first case it possible to apply the technique of time-correlated single photon counting to exciton kinetic measurements under short laser pulse excitation. This is to be contrasted with the second case where spectral measurements and molecular jumps can be studied under CW laser excitation. In both cases it is possible to apply microscopy methods.

3.17.2 Combination of TCSPC and NSOM: New Advantages

At the Lebedev Physical Institute in Moscow, a unique setup has been installed. The new setup combines the facilities of two techniques: NSOM and TCSPC. The TCSPC system is based on a femtosecond Ti–Sapphire laser pumped by an Ar ion laser with T-format online detection.

In this way, doped or pure thin film luminescence kinetics can be studied with maximal for optics limit. It gives information about exciton dynamics between nanosize species or between macromolecule fragments. This method is extremely important for studying the above-mentioned antenna and shell effects associated with dendrimers.

3.17.3 Kinetics of Energy Transfer Between the NSOM Tip and Single Molecules/Ensembles on Surfaces

A new microscopic technique [43] is demonstrated that combines attributes from both near-field scanning optical microscopy (NSOM) and fluorescence resonance energy transfer (FRET). The method relies on attaching the acceptor dye of a FRET pair to the end of a near-field fiber optic probe (tip). Light exiting the NSOM probe, which is non-resonant with the acceptor dye, excites the donor dye introduced into a sample. As the tip approaches the sample containing the donor dye, energy transfer from the excited donor to the tip-bound acceptor produces a redshifted fluorescence.

The main information about FRET can be taken from the fluorescence kinetics of the donor or acceptor. Hence in this case a combination of NSOM and correlated single photon counting methods is needed.

This method is shown to provide enhanced depth sensitivity in fluorescence measurements, which may be particularly informative in studies on thick specimens such as cells. The technique also provides a mechanism for obtaining high spatial resolution without the need for a metal coating around the NSOM probe and should work equally well with non-waveguide probes such as atomic force microscopy tips. This may lead to dramatically improved spatial resolution in fluorescence imaging.

Acknowledgements

AGV would like to thank Prof. M. Van der Auweraer, Prof. A. Muzafarov, and Dr. I. Scheblykin for their cooperation and joint research and Prof. R. Silbey and Prof. R. Kopelman for fruitful discussions. Financial support was made available through NATO SfP 97-1940, and RFBR grants 99-02-17326 and 00-15-96707. ISO wishes to thank RFBR for financial support through grant 01-02-16580.

References

1. G. Binning, H. Rohrer, C. Gerber, E. Weibel: Phys. Rev. Lett. **50**, 120 (1983)
2. G. Binning, H. Rohrer: Angew. Chem. Int. Ed. Eng. **26**, 606 (1987)
3. *Single Charge Tunneling. Coulomb Blockade Phenomena in Nanostructures*, ed. by H. Grabert and M.H. Devoret, NATO ASI Series, Series B: Physics, vol. 294 (1992), Plenum Press, NY, London
4. R. Berndt, R. Gaisch, W. Schneider, J. Gimzewski, B. Reihl, R. Schlittler, M. Tschudy: Phys. Rev. Lett. **74**, 102 (1995)
5. W.E. Moerner, L. Kador: Phys. Rev. Lett. **62**, 2535 (1989)
6. M. Orrit, J. Bernard: Phys. Rev. Lett. **65**, 2716 (1990)
7. S.A. Soper, E.B. Shera, J.S. Martin, J.H. Jett, J.H. Hahn, H.J. Nutter, R.A. Keller: Anal. Chem. **63**, 432 (1991)
8. *Single Molecule Optical Detection, Imaging and Spectroscopy*, ed. by T. Basche, W.E. Moerner, M. Orrit, U. Wild (VCH, Weinheim 1996)
9. P. Thamarat, A. Maali, B. Lounis, M. Orrit: J. Phys. Chem. A **104**, 1 (2000)
10. M. Paesler, P. Moyer: *Near-Field Optics* (John Wiley, New York 1996)
11. T. Basche, W.E. Moerner, M. Orrit, H. Talon: Phys. Rev. Lett. **69**, 1516 (1992)
12. R. Berndt: In: *Scanning Probe Microscopy*, ed. by R. Wiesendanger (Springer, Berlin, Heidelberg, New York 1998)
13. S.A. Kulagin, I.S. Osad'ko: Phys. Stat. Sol. (B) **110**, 184 (1982)
14. I.S. Osad'ko: In: *Spectroscopy and Excitation Dynamics of Condensed Molecular Systems*, ed. by V. Agranovich and R. Hochstrasser (North-Holland, Amsterdam 1983)
15. C. Joachim, J. Gimzewski, R. Schlittler, C. Chavy: Phys. Rev. Lett. **74**, 2102 (1995)

16. M. Cuberes, R. Schlittler, J. Gimzewski: Appl. Phys. Lett. **69**, 3016 (1996)
17. C. Joachim, J. Gimzewski, H. Tang: Phys. Rev. B **58**, 16407 (1998)
18. T. Jung, R. Schlittler, J. Gimzewski: Nature **386**, 696 (1997)
19. D. Fujita, T. Ohgi, W.-E. Deng, H. Nejo, T. Okamoto, S. Yokoyama, K. Kamikado, S. Mashiko: Surf. Sci. **454–456**, 1021 (2000)
20. S. Iakovenko, A. Trifonov, S. Soldatov, V. Khanin, S. Gubin, G. Khomutov: Thin Solid Films **284–285**, 873 (1996)
21. J. Gimzewski, J. Sass, R. Schlittler, J. Schott: Europhys. Lett. **8**, 435 (1989)
22. R. Berndt, J. Gimzewski, P. Johansson: Phys. Rev. Lett. **67**, 3796 (1991)
23. O. Keller: Phys. Lett. A **188**, 272 (1992)
24. I. Smolyaninov: Surf. Sci. **364**, 79 (1996)
25. P. Johansson, R. Monreal, P. Apell: Phys. Rev. A **14**, 9210 (1990)
26. L. Landau, E. Lifshitz: *Field Theory* (Fizmatgiz, Moscow 1960)
27. M. Xiao: Phys. Rev. Lett. **82**, 1875 (1999)
28. P. Johansson, R. Berndt, J. Gimzewski, S. Apell: Phys. Rev. Lett. **84**, 2034 (2000)
29. M. Xiao: Phys. Rev. Lett. **84**, 2035 (2000)
30. I.S. Osad'ko, L.B. Yershova: J. Lumin. **87–89**, 184 (2000)
31. I.S. Osad'ko, L.B. Yershova: J. Lumin. **86**, 211 (2000)
32. S.A. Soper, L.M. Davis, E.B. Shera: J. Opt. Soc. Am. **B9**, 1761 (1992)
33. S. Kullback: *Information, Theory and Statistics* (Wiley, New York 1959) Sect. 6.4
34. J. Enderlein, P.M. Goodvin, A.V. Orden, W.P. Ambrose, R. Erdmann, R.A. Keller: Chem. Phys. Lett. **270**, 454 (1997)
35. M. Kollner, S. Wolfrum: Chem. Phys. Lett. **200**, 199 (1992)
36. M. Kollner, A. Fischer, J. Arden-Jacob, K.H. Drexhage, R. Muller, S. Seeger, J. Wolfrum: Chem. Phys. Lett. **250**, 355 (1996)
37. M. Kollner: Appl. Opt. **32**, 501 (1993)
38. C. Zander, M. Sauer, K.H. Drexhage, D.-S. Ko, A. Schuiz, J. Wolfrum, C. Eggeling, C.A.M. Seidel: Appl. Phys. B **63**, 517 (1996)
39. E.B. Shera, N.K. Seitzinger, L.M. Davis, R.A. Keller, S.A. Soper: Chem. Phys. Lett. **174**, 553 (1990)
40. S. Tretak, V. Chernyak, S. Mukamel: J. Chem. Phys. B **102**, 3310 (1998)
41. R. Kopelman, M. Shortreed, Z.Y. Shi, W. Tang, Z. Xu, J. Moore, A. Bar-Haim, J. Klafter: Phys. Rev. Let. **78**, 1239 (1997)
42. G.M. Stewart, M.A. Fox: J. Am. Chem. Soc. **118**, 4354 (1996)
43. P.W. Wang, Y.J. Liu, C. Devadoss, P. Bharathi, J.S. Moore: Adv. Mater. **8**, 237 (1996)
44. A.G. Vitukhnovsky, M.I. Sluch, V.G. Krasovskii, A.M. Muzafarov: Synthetic Metals **91**, 375 (1997)
45. M.I. Sluch, I.G. Scheblykin, O.P. Varnavsky, A.G. Vitukhnovsky, V.G. Krasovskii, O.B. Gorbatsevich, A.M. Muzafarov: J. Luminescence **76–77**, 246 (1998)
46. D.A. Tomalia, A.M. Naylor, W.A. Goddard III: Angew. Chem. Int. Ed. Engl. **29**, 138 (1990)
47. A. Archut, F. Vogtle: Chem. Soc. Reviews **27**, 233 (1998)
48. R.G. Denkewaller, J.F. Kolc, W.J. Lukasavage: US Patent 4.410.688 (1983)
49. G.R. Newkome, C.N. Moorefield, G.R. Baker: Aldrichchimica Acta **25**, 31 (1992)

50. K.L. Wooley, C.J. Hawker, J.M. Frechet: J. Am. Chem. Soc. **113**, 4252 (1991)
51. D.A. Tomalia: Adv. Mater. **6**, 529 (1994)
52. E.E. Jelley: Nature **138**, 1009 (1936)
53. G. Scheibe: Angew. Chem. **49**, 563 (1936)
54. S. de Boer and D.A. Wiersma: Chem. Phys. Lett. **165**, 45 (1990)
55. F. Spano, J.R. Kuklinski, S. Mukamel: Phys. Rev. Lett. **65**, 211 (1990)
56. H. Fidder, D.A. Wiersma: Phys. Stat. Sol. (B) **188**, 285 (1995)
57. V. Kamalov, I.A. Struganova, K. Yoshihara: J. Chem. Phys. **100**, 8640 (1996)
58. S. Kobayashi, F. Sasaki: Nonlinear Optics **4**, 305 (1993)
59. V. Sundström, T.Gillbro, R.A. Gadonas, and A. Piskarskas: J. Chem. Phys. **89**, 2754 (1988)
60. R. Kopelman: In: *Spectroscopy and Excitation Dynamics of Condensed Molecular Systems*, ed. by V.M. Agranovich and R.M. Hochstrasser (North Holland, Amsterdam 1983)
61. V. Czikkley, H.D. Foerstering, H. Kuhn: Chem. Phys. Lett. **6**, 11 (1970)
62. H.J. Nolte: Chem. Phys. Lett. **31**, 134 (1975)
63. S. Kirstein, H. Moevald: Advanced Materials **7**, 460 (1995)
64. D.A. Higgins, P.J. Reid, P.F. Barbara: J. Phys. Chem. **100**, 1174 (1996)
65. H. Fidder and D.A. Wiersma: J. Phys. Chem. **97**, 11603 (1993)
66. D.L. Smith: Photogr. Sci. Eng. **18**, 309 (1974)
67. A.E. Johnson, S. Kumazaki, K. Yoshihara: Chem. Phys. Lett. **211**, 511 (1993)
68. B. Kopainsky, J.K. Hallermaeier, and W. Kaiser: Chem. Phys. Lett. **83**, 498 (1981)
69. I.G. Scheblykin, M.A. Drobizhev, O.P. Varnavsky, M. Van der Auweraer, A.G. Vitukhnovsky: Chem. Phys. Lett. **261**, 181 (1996)
70. T. Kobayashi, K. Misawa: In: *Excitonic Processes in Condensed Matter EXCON'96*, Proc. of the 2nd International Conference, Kurort Gohrisch, Germany, August 14–17, 1996, ed. by M. Schreiber (Dresden, University Press 1996) pp. 23–26
71. M.A. Drobizhev, M.N. Sapozhnikov, I.G. Scheblykin, O.P. Varnavsky, M. van der Auweraer, and A.G. Vitukhnovsky: Chem. Phys. **211**, 456 (1996)
72. H. Hada, C. Honda, H. Tanemura: Photogr. Sci. Eng. **21**, 83 (1977)
73. E. Rap, M. Ketelaars, J. Borst, A. van Hoek, A. Visser: Biophys. Chem. **58**, 255 (1996)
74. M.I. Sluch, I.G. Scheblykin, O.P. Varnavsky, A.G. Vitukhnovsky, V.G. Krasovskii, O.B. Gorbatsevitch, A.M. Muzafarov: J. luminescence **76–77**, 246 (1998)
75. V.G. Krasovskii, N.A. Sadovskii, O.B. Gorbatsevitch, A.M. Muzafarov, V.D. Myakushev: Polym. Sci. **36**, 589 (1994)
76. J. Feldmann, G. Peter, E.O. Gobel, P. Dawson, K. Moore, C.Foxon, R.J. Elliot: Phys. Rev. Lett. **59**, 2337 (1987)
77. D. Mobius, H. Kuhn: Isr. J. Chem. **18**, 375 (1979)
78. E.O. Potma, D.A. Wiersma: J. Chem. Phys. **108**, 4894 (1998)
79. S. De Boer, D.A. Wiersma: Chem. Phys. Lett. **165**, 45 (1990)
80. H. Fidder, D.A. Wiersma: Phys. Status Solidi B **188**, 285 (1995)
81. V.F. Kamalov, l.A. Struganova, K. Yoshihara: J. Phys. Chem. **100**, 8640 (1996)
82. J. Moll, S. Daehne, J.D. Durrant, D.A. Wiersma: J. Chem.Phys. **102**, 6362 (1995)

83. M. Lindrum, A. Glismann, J. Moll, S. Daehne: Chem. Phys. **178**, 423 (1993)
84. U. De Rossi, U. Stahl, S. Dahne, M. Lindrum, S.C.J. Meskers, H.P.J.M. Dekkers: J. Fluores. **7**, 715 (1997) Supplement
85. P. Esquinazi (Ed.): *Tunneling Systems in Amorphous and Crystalline Solids* (Springer, Berlin, Heidelberg, New York 1998)
86. I. Osad'ko, L. Yershova: J. Chem. Phys. **111**, 7652 (1999)
87. W.E. Moerner,T. Basche: Angew. Chem. Int. Ed. Engl. **32**, 457 (1993)
88. W.P. Ambrose, T. Basche, W.E. Moerner: J. Chem. Phys. **95**, 7150 (1991)
89. P. Thenio, A.B. Myers, W.E. Moerner: J. Lum. **56**, 1 (1993)
90. L. Fleury, A. Zumbusch, M. Orrit, R. Brown, J. Bernard: J. Lum. **56**, 15 (1993)
91. A.M. Boiron, P. Tamarat, B. Lounis, R. Brown, M. Orrit: Chem. Phys. **247**, 119 (1999)
92. H.P. Lu, X.S. Xie: Science **282**, 1877 (1998)
93. I.S. Osad'ko: JETP **86**, 875 (1998)
94. I.S. Osad'ko: JETP **89**, 513 (1999)
95. I.S. Osad'ko, L.B. Yershova: J. Chem. Phys. **112**, 9645 (2000)
96. P.W. Anderson, B.I. Halperin, C.M. Varma: Philos. Mag. **25**, 1 (1972)
97. W.A. Phillips: J. Low Temp. **7**, 352 (1972)
98. K.P. Muller, D. Haarer: Phys. Rev. Lett. **66**, 2344 (1991)
99. I.S. Osad'ko: Adv. Polym. Sci. **114**, 123 (1994)
100. H. Talon, L. Fleury, J. Bernard, M. Orrit: JOSA B **9**, 825 (1992)
101. B. Kozankevich, J. Bernard, J. Orrit: J. Chem. Phys. **101**, 9377 (1994)

4 Scanning Tunneling Spectroscopy and Electronic Properties of Single Fullerene Molecules

J.G. Hou and B. Li

The discovery of an effective preparation procedure for obtaining macroscopic quantities of C_{60}, a third form of carbon besides graphite and diamond, has triggered a significant research effort to understand the physical and chemical properties of C_{60} and C_{60}-related fullerene molecules [1]. Fullerene molecules are closed cage molecules containing only hexagonal and pentagonal faces. Chemically stable, cage-structured C_{60} molecules form an interesting new family of adsorbates on surfaces. They differ significantly from the elemental or simple molecular adsorbates because of their three-dimensional character on the atomic scale. A unique fundamental property of this type of adsorbate is molecular orientation with respect to the host substrate. When an isolated molecule ceases its rotational motion on a surface, it may, in general, adopt a number of binding configurations and hence a range of different orientations. Interaction between molecules could yield still other orientational arrangements. Understanding site- or orientation-dependent C_{60}–substrate interactions therefore lies at the heart of the design of new catalysts with functionalized cage molecules or the fabrication of thin films with desirable orientational orders. However, most surface analysis techniques (e.g., HREELS, PES) can only provide statistically averaged results rather than spatially resolved results.

On the other hand, device sizes will ultimately decrease down to the nanometer scale and the building blocks and functional units of nano-devices will be individual molecules and even individual atoms in the near future. Recent breakthroughs have been achieved through scientists' skill in manipulating individual atoms and molecules with STM, and a variety of molecular devices and concepts like atom relays, single-atom switches and molecular amplifiers have been proposed and realized [2–6]. Cage-structured C_{60} molecules possess many interesting physical and chemical properties and have shown their amazing potential in applications for molecular devices. For example, the first single-molecule electromechanical amplifier with a gain of five has been produced by a metal tip pressing a C_{60} molecule [6]. This is based on the modulation of virtual resonance tunneling through the C_{60} molecule by electromechanical deformation of the C_{60} cage. The interplay between the single-electron tunneling effect and splitting of degenerate molecular levels has also been observed for C_{60} molecules deposited on an insulating layer at 4.2 K [7].

In this chapter, the method of scanning tunneling spectroscopy (STS) is introduced first and then recent STM/STS studies on C_{60} molecules are reviewed. We will focus on the following two aspects:

- local electronic structures of individual C_{60} molecules on Si substrates,
- electronic transport properties of individual C_{60} molecules.

4.1 Scanning Tunneling Spectroscopy

Scanning tunneling microscopy (STM) was invented by G. Binnig and H. Rohrer [8–10]. The basic principle of the instrument is quantum mechanical tunneling of electrons between a metallic tip and the sample which can be a flat surface of any conducting or semiconducting material. The magnitude of the tunneling current is exponentially proportional to the inverse distance between the tip and the surface, so it is possible to obtain atomic resolution images by scanning the tip over the surface. If the tip is held over a given point on the surface of the sample, and the voltage is swept to make a series of current measurements as a function of voltage, an I–V curve can then be measured corresponding to that location. According to the theory of Tesoff and Hamann [11], the tunneling current in the STM can be expressed as

$$I(V) \propto \int_{E_\mathrm{F}}^{E_\mathrm{F}+eV} \rho(\boldsymbol{r}, E)\,\mathrm{d}E\,, \tag{4.1}$$

$$\rho(\boldsymbol{r}, E) = \sum_i |\psi_i(\boldsymbol{r})|^2\,\delta(E - E_i)\,, \tag{4.2}$$

where $\rho(\boldsymbol{r}, E)$, $\psi_i(\boldsymbol{r})$, and E_F are the local density of states (LDOS) of the sample, the sample wave function with energy E_i, and the Fermi energy, respectively. The normalized derivative $(\mathrm{d}I/\mathrm{d}V)/(I/V)$ of this curve, where I is the tunneling current and V the sample bias voltage, reflects the LDOS of the samples in a first order approximation [12].

Consequently, scanning tunneling spectroscopy (STS) was expected to offer a novel view of the electronic structure of surfaces because it could probe the energy dependence of the sample electronic states with unprecedented spatial resolution. Although early studies fulfilled this promise in the dangling-bond semiconductor [13], the progress of STS in metals has been slower than expected. The difficulties encountered arise on both the experimental and the theoretical sides. The results for the tunneling current as well as the spectroscopy depend on the geometric and electronic structures of the tip. A blunt tip with a featureless LDOS is required for reproducible spectroscopy reflecting the LDOS of the sample in the first approximation mentioned above [12]. But a blunt tip is easily 'sharpened' by accidentally attaching a foreign adatom or small cluster of atoms during the scanning. Although this is good for producing a high resolution image, such sharpening

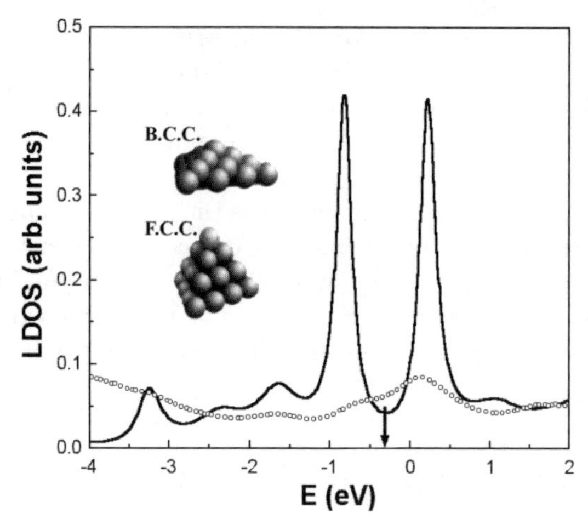

Fig. 4.1. Calculated local density of states (s states) at the apex atom for bcc (*line composed of circles*) and fcc (*solid line*) (111)-oriented W pyramids. The *arrow* indicates the position of the Fermi level. The *inset* shows the two tip geometries considered [14]

leads to a tip structure exhibiting resonant electronic states that strongly modify the spectra. On the other hand, even the clean tip may have resonant states around the Fermi energy if the tip ends with a specific structure. For example, Vázquez et al. [14] have shown both experimentally and theoretically that while the (111) bcc W-tip yields a smooth, bulklike density of states (DOS), the (111) fcc tip exhibits two narrow resonances within the upper energy range of the bulk spectrum on both sides of E_F (see Fig. 4.1). Despite the lack of complete experimental control over the tip structure and consequent difficulties of reproducibility in the practical context, STS is still a truly unique way to probe local electronic properties.

Recently, applications of STM to molecules or superamolecular assemblies have attracted considerable interest. The STS of individual adsorbed molecules is expected to provide valuable information about the electronic structure and quantum transport properties. However, this information depends on the molecule–substrate interaction and coupling of the molecule to its environment. This is therefore a challenging task and several key issues have to be addressed in addition to the problems for semiconductor and metal surfaces. Firstly, STS measurements for molecules are usually conducted on the molecule–substrate system, so suitable atomically flat and conducting substrates have to be found. Secondly, interpretation of the tunneling spectrum of molecular species requires a deep understanding of their electronic structure and transport properties. If the interaction between molecule and substrate is strong, it will be more difficult to understand the peaks in the spectrum since the interaction gives rise to strong chemisorption bonds and a corresponding alteration of the energy levels of the adsorbed molecules. Thirdly, it is important to eliminate the surface mobility of the molecules to allow stable STM imaging and spectrum measurement. This may be achieved in different ways, for instance, by physisorption at low temperatures or self-organization of molecular structures.

In the following sections, we will use the C_{60} molecule as the model system in order to show how to obtain information on molecular electronic structure, site- and orientation-dependent molecule–substrate interactions and electronic transport properties from elaborate STS experiments combined with theoretical calculations. We will also discuss single-molecule devices using C_{60} as functional unit.

4.2 Electronic Structures of Fullerene Molecules

The structure model of the C_{60} molecule was proposed by Kroto et al. in 1985 [15]. It consists of twelve pentagonal rings and twenty hexagonal rings, and no pentagonal rings make direct contact each other (see Fig. 4.2). According to NMR experiments [16], all C atoms are identical in a C_{60} molecule. The bond lengths are 1.46 Å and 1.40 Å, respectively, for the C–C single bonds between a pentagon and a hexagon and C=C double bonds between two hexagons. The bonding character of C_{60} is predominantly sp^2 with a small admixture of sp^3 character due to the non-zero curvature. For this reason, the electronic states can be decomposed into approximate π and σ states. The bonding σ states reside well below the highest occupied level, which is composed of orbitals having primarily π character [17,18]. The nearly spherical structure of the C_{60} suggests a labelling of these electronic states in terms of spherical harmonics, with the σ and π electrons corresponding to different radial quantum numbers. Theoretical calculations [19] for the isolated C_{60} molecule show the relative positioning in energy of these molecular orbitals (Fig. 4.3). The π electrons pair up to fill 30 energy states. The degeneracy of the angular momentum $L = 5$ level, which contains 11 states, is removed by the icosahedral symmetry of the molecule, resulting in three separate levels:

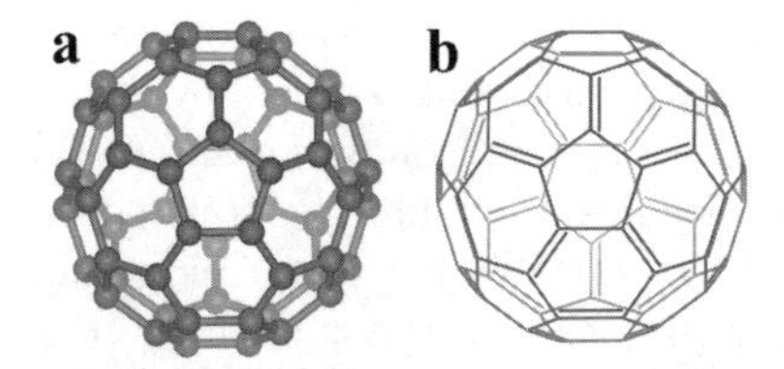

Fig. 4.2. Ball-and-stick (**a**) and bond model (**b**) of a C_{60} molecule. Each bond on the pentagonal rings is a single C–C bond, while those shared by two neighboring hexagonal rings are double C=C bonds

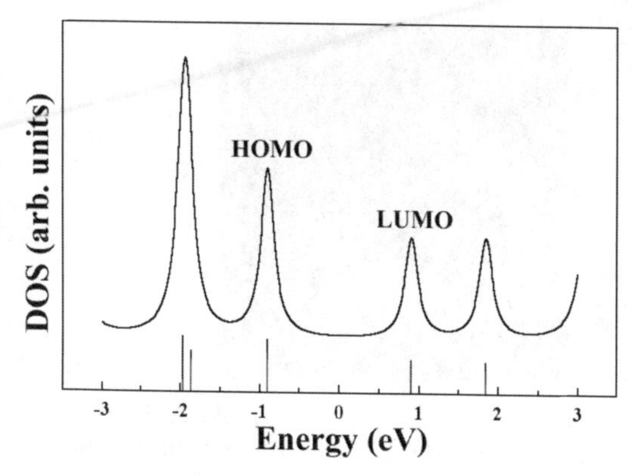

Fig. 4.3. Density of states for the isolated C_{60} molecule calculated using the discrete variational LDA method. *Vertical lines* denote energy levels of the molecule. The DOS is obtained by broadening the discrete levels with a Lorentz function for better comparison with experimental data

- the h_u, or highest occupied molecular orbital (HOMO), containing 10 electrons,
- the triply degenerate t_{1u}, or lowest unoccupied molecular orbital (LUMO), which can accommodate 6 electrons,
- a similar triply degenerate t_{2u} orbital (the LUMO +1) at higher energy.

In order to study the electronic structure of an isolated C_{60} molecule using STS, it is necessary to put the molecules onto a substrate that has little or no interaction with C_{60}. A C_{60} monolayer supported on a self-assembled monolayer (SAM), as illustrated in Fig. 4.4a, is a good structure because the SAM is atomically flat and interacts very weakly with the C_{60} molecules. The preparation of the SAM substrate takes two steps. A 160 nm thick Au(111) film is first grown on a freshly cleaved and well-outgassed mica sheet by vacuum deposition at 300°C. The mica-supported gold film is then quenched in ethanol and, while still covered with an ethanol droplet, immersed in a 2 mM ethanol solution of alkylthiol ($CH_3(CH_2)_8SH$) for about 48 hours at room temperature. This procedure routinely yields high quality SAM as verified by STM. A piece of the SAM substrate is cut and transferred into the ultra-high vacuum and low-temperature STM chamber where a submonolayer of C_{60} is then deposited onto the substrate. The STM investigation shows that the C_{60} molecules form close-packed hexagonal arrays on the SAM substrate (see Fig. 4.4b).

At room temperature, the C_{60} arrays are unstable. Molecules at the edge of an array can detach readily and diffuse to another part of the same array or other nearby arrays. This observation is truly indicative of a very weak interaction between the C_{60} and the alkylthiol. The STS measurements were performed at 78 K using an UHV LT-STM with a Pt–Ir tip which was subjected to a careful cleaning treatment. Figure 4.5 displays a typical DOS that

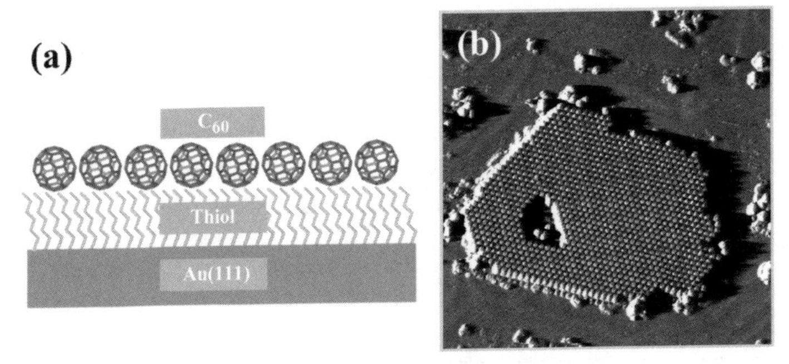

Fig. 4.4. (**a**) Schematic of a C_{60} overlayer supported on the thiol SAM. (**b**) STM image of C_{60} close-packed hexagonal arrays on the SAM substrate ($40 \times 40\,\mathrm{nm}^2$)

was derived from I–V spectra measured by positioning the tip above an inner molecule in a selected island. The low temperature and the fact that C_{60} molecules are assembled together both ensure that the molecules do not move around during tip scanning and spectrum measurement. Note that adjacent C_{60} molecules do not modify each other's electronic structures due to van der Waals intermolecular interactions, so STS data is expected to contain information relevant to an isolated molecule. Indeed, we find from Fig. 4.5 that the main features of the experimental results are in good agreement with calculated theoretical results.

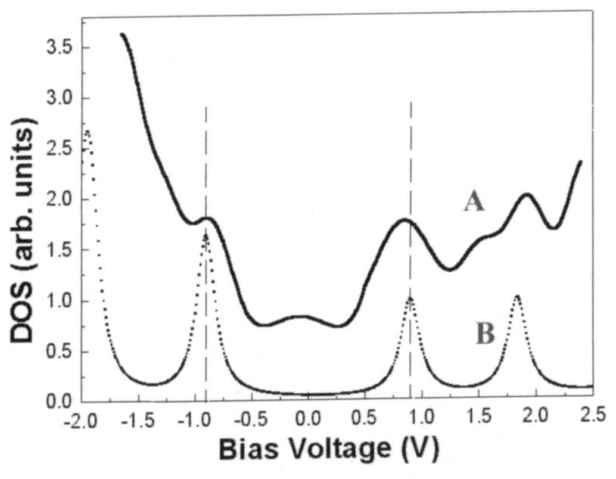

Fig. 4.5. Density of states (A) for a C_{60} molecule in a close-packed hexagonal island on alkylthiol (derived from the experimental I–V curve). Curve B is the theoretically calculated DOS for the isolated C_{60} molecule

4.3 Scanning Tunneling Spectroscopy of Single C_{60} Molecules Adsorbed on a Si Surface

It is very important to study fullerene–surface interactions if we are to understand the fundamental physical and chemical properties of fullerene and develop potential applications. The adsorption characteristics and molecule–substrate interaction of C_{60} on semiconductor surfaces as well as on metal surfaces have been investigated extensively by STM [20] and other techniques such as high-resolution electron-energy-loss spectroscopy (HREELS) [21,22] and photoelectron spectroscopy (PES) [23–25]. In this section we will discuss the scanning tunneling spectroscopy of single C_{60} molecules on an Si surface, and focus on local electronic properties that depend on the adsorption site and molecular orientation.

4.3.1 C_{60} Molecules on an Si(111)–(7×7) Surface

The structure of the Si(111)–(7 × 7) surface has been extensively studied and is now well established. According to the accredited model, Si atoms take several different bonding configurations within the 7 × 7 unit cell, and each corner hole atom, adatom and rest atom has one dangling bond for the Si(111)–(7 × 7) surface. A strong interaction between the silicon surface and fullerenes could therefore be expected due to the high density of dangling bonds. Early STM studies of C_{60} on Si(111) indicated that there are four different adsorption sites (see Fig. 4.6) when C_{60} coverage is much less than monolayer, and STM images showed similar strip-like intramolecular structures with positive bias voltage [26]. Moreover, Li et al. [27] found that the height of C_{60} is strongly dependent on the bias voltage. In principle, the pentagonal and hexagonal faces or edge atoms of the C_{60} molecule could bond differently with dangling bonds on surface Si atoms, and this results in various bias-dependent characteristics. Apart from showing that the interaction

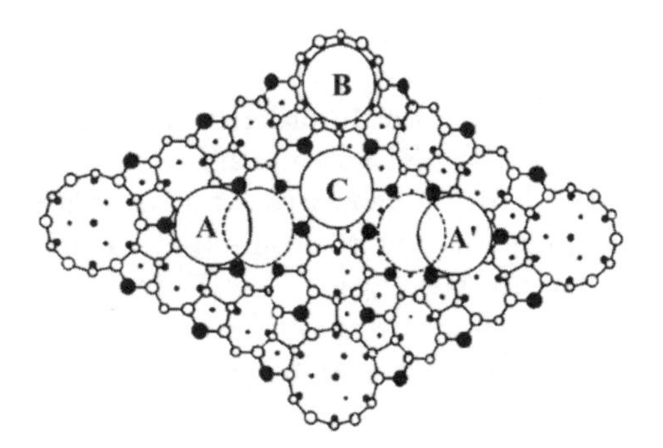

Fig. 4.6. Schematic of four possible C_{60} adsorption sites on Si(111)–(7×7): A faulted half, A′ unfaulted half, B corner holes, C dimer lines. About 80% of the C_{60} molecules are on sites A and A′, 13% on site B, and 7% on site C

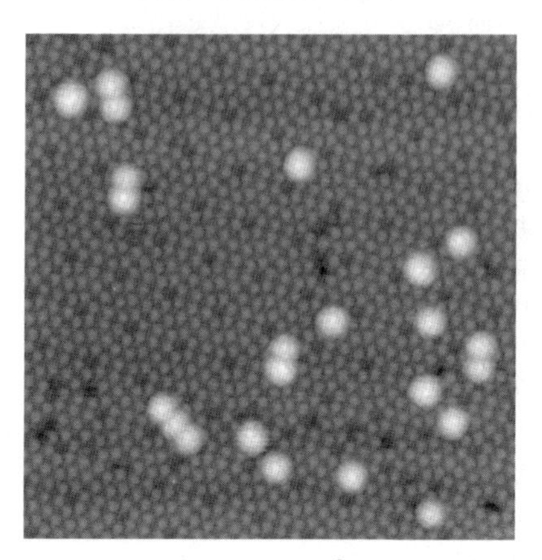

Fig. 4.7. A $25 \times 25\,\text{nm}^2$ STM topography image of 0.05 monolayer C_{60} deposited on an Si(111)–(7×7) surface

between adsorbed C_{60} molecules and substrate is strong, and that differences in bonding configuration (or orientation) exist for individual C_{60} molecules on the different sites, these experiments have left the challenge of identifying the specific molecular orientation and its local electronic structure largely unanswered.

Recently, we used low temperature STM/STS to identify the molecular orientation of individual C_{60} molecules and probed their local electronic structure on Si(111)–(7×7) surfaces [28,29].

The experiments were carried out using LT-STM with an electrochemically etched tungsten tip. All STM topographies and STS were obtained at 78 K in order to reduce thermal noise. The tip–sample separations for STS measurements were controlled by setting the sample bias voltage at 2.0 V and setting the tunneling current in the range 0.1–0.2 nA. Figure 4.7 shows a $25 \times 25\,\text{nm}^2$ STM topography image of 0.05 monolayer C_{60} deposited on an Si(111)–(7×7) surface.

Curves A, B and C in Fig. 4.8 are typical STS results corresponding to C_{60} adsorbed at sites A, B, C, respectively, on an Si(111)–(7×7) surface. Curve D is the average STS result on the bare Si surface. In order to discuss the energy band structure of the adsorbed C_{60}, these STS results are represented in a graph of $(\mathrm{d}I/\mathrm{d}V)/(I/V)$ vs. V. It is already known that the tip status is of great importance in STS investigations. In order to ensure that our STS results are reliable and reproducible, we use the Si(111)–(7×7) surface as a reference. Before and after the STS measurements on a C_{60} molecule, we performed STS measurements on the Si surface. The results indicate that the STS results on the Si surface are reproducible and have the same features as reported previously [13].

We found that the band gaps of C_{60} adsorbed at sites A, B and C are 1.4, 0.8 and 1.3 eV respectively. These values are much smaller than the value for

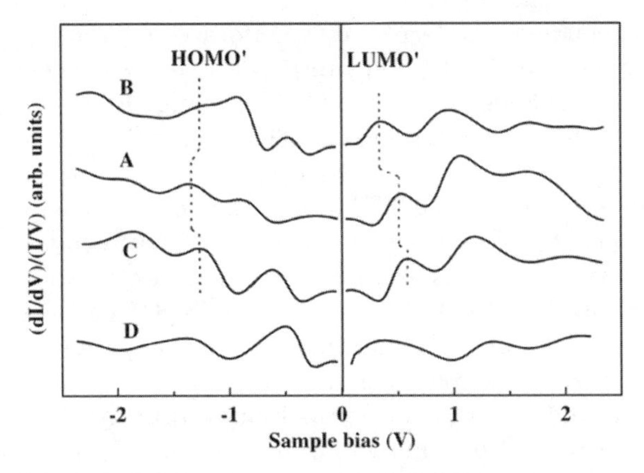

Fig. 4.8. STS results: curves A, B and C are the averaged STS results corresponding to C_{60} adsorbed at sites A, B, and C, respectively, on an Si(111)–(7 × 7) surface. Curve D is the averaged STS result on the Si(111)–(7 × 7) surface

free C_{60} (1.9 eV, see Fig. 4.2 and [30,31]). Previous studies have shown that C_{60} molecules adsorbed on an Si(100)–(2 × 1) surface have a band gap of about 1.8 eV [33,34]. The reduced band gap of C_{60} adsorbed on the Si(111)–(7 × 7) surface can be explained as follows. Because of the strong interaction between C_{60} and dangling bonds of the Si(111)–(7×7) surface, C_{60} molecules are distorted by bonding with Si atoms and their symmetry is reduced. The splitting orbitals derived from the symmetry reduction will mix with the surface states of the Si(111)–(7 × 7) surface [35,24]. In this way, the LDOS of the Si surface is incorporated into the band gap of the C_{60} molecules. Therefore, strictly speaking, the STS results in Fig. 4.8 no longer originate from C_{60} molecules, but from a 'compound' made from a C_{60} molecule and Si atoms. For convenience, we will still denote these curves as the STS of C_{60} molecules in the following unless otherwise specified.

Moreover, since the surface structures of Si(111)–(7 × 7) differ from one another at different sites, so the mixing process of the C_{60} orbitals with the LDOS of Si(111)–(7 × 7) is also site-dependent. It is observed that the STS characteristics of sites A, B, and C are more similar at unoccupied states than at occupied states. This suggests that occupied orbitals of the C_{60} molecule mix more strongly than its unoccupied orbitals with Si surface states.

The HOMO and LUMO are the orbitals that dominate the physical and chemical properties of adsorbed C_{60} molecules, so it is important to identify them in our STS results. Even though as discussed above the interaction between the C_{60} molecule and the Si surface is strong, it is still reasonable to expect the original orbitals of free C_{60} molecules to remain traceable. The reason for this is that C_{60} molecules are rather large and stable. The cage structure of the C_{60} molecule cannot be changed, so distortion of the C_{60} molecules should be small. Moreover, the interaction between C_{60} and Si surface is localized at the bonding interface. Taking into account the fact that the original band gap between HOMO and LUMO for free C_{60} is about 1.6–1.9 eV [30,31] and comparing STS curves with free C_{60} energy levels, the HOMO and LUMO of C_{60} molecules are assigned to the peaks of the STS as labeled in Fig. 4.8.

To obtain a deeper understanding of the STS results and verify the accuracy of the above discussion concerning electronic properties of adsorbed C_{60} molecules, further theoretical calculations are necessary. However, we must first identify molecular orientations from the high resolution STM images before carrying out any calculation. It was found that the internal pattern of the C_{60} depends strongly on the bias voltage and the tip–sample distance. Figure 4.9 shows the bias dependent results of two molecules adsorbed on an A-site (Figs. 4.9a1, a2) and a B-site (Figs. 4.9b1, b2), respectively. When the sample is positively biased, i.e., electrons tunnel from the tip to unoccupied molecular orbitals and empty states of the substrate, an intramolecular pattern with one bright pentagon ring plus two curved strokes is observed at the A-site. Similar reproducible results were also obtained for C_{60} molecules on A′-sites. This is not surprising since the binding configurations are similar for these two sites (ignoring the faulted/unfaulted sublayers). At an A-site (or A′-site), the molecule appears as four bright stripes (Fig. 4.9a1) when the sample is biased at -1.8 V with respect to the tip, and the electrons tunnel from the occupied molecular orbitals and the filled states of the substrate to the tip. On site B, Fig. 4.9b2 shows that the internal pattern of the C_{60} at positive sample bias is a bright pentagon ring plus three curved strokes to its right, which is significantly different from the pattern on site A (Fig. 4.9a2). By contrast, Fig. 4.9b1 shows that the negative bias STM image on site B is quite similar to that on site A (Fig. 4.9a1), both exhibiting four slightly curved bright stripes.

Reproducible high resolution STM images, such as Figs. 4.9a1, 4.9a2, 4.9b1 and 4.9b2, showing clear intramolecular features of the C_{60} molecules, raise the possibility of identifying the molecular orientations on Si(111)–(7×7) surfaces. In general, however, the observed internal structures are not directly related to the internal atomic configurations, but rather to the electronic structure of the adsorbed fullerenes and surfaces. Mapping from the latter to the former clearly entails theoretical modeling.

We adopted Tersoff and Hamann's formula [11] and its extension in order to simulate STM images. Equations (4.1) and (4.2) assume a constant density of states of the tip, which allows us to obtain STM images only from the LDOS of the sample surface. This assumption is tenable when the separation between tip and sample is big enough. The experimental conditions for obtaining Figs. 4.9a1, 4.9a2, 4.9b1 and 4.9b2 meet this requirement. It is known that the LDOS near E_F makes a dominant contribution to STM images, so we can obtain the LDOS by adding up several orbitals around the highest

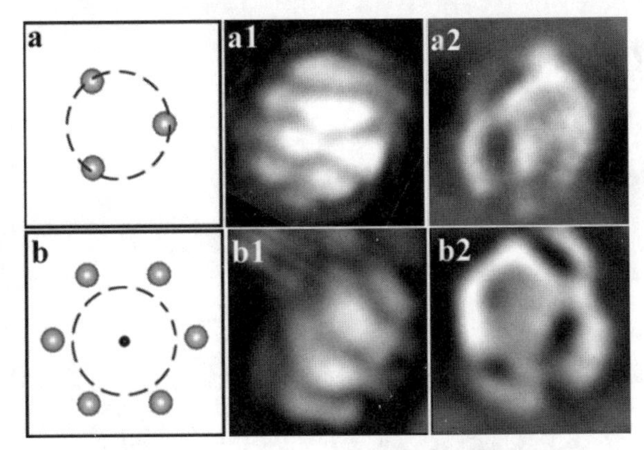

Fig. 4.9. High resolution STM images of individual C_{60} molecules recorded with different bias voltages on an A-site of the Si(111)–(7×7) surface: (**a1**) $V_s = -1.8$ V, (**a2**) $V_s = 2.5$ V. On a B-site: (**b1**) $V_s = -1.8$ V, (**b2**) $V_s = 2.3$ V. The tunneling currents are 0.15 nA for positive bias images and 0.1 nA for negative bias images

occupied molecular orbital and the lowest unoccupied molecular orbital with the same weight.

We used cluster models to mimic an individual C_{60} molecule adsorbed on Si(111)–(7 × 7) with a number of binding configurations. In these cluster models, the Si dangling bonds irrelevant to C_{60} adsorption are saturated by hydrogen atoms. We only considered two adsorption sites A and B. For C_{60} adsorbed on site A, our cluster model is $C_{60}Si_{57}H_{42}$, while it is $C_{60}Si_{67}H_{45}$ for the B site. Figure 4.10 shows schematic views of the cluster models.

The electronic structures of the clusters were calculated using the discrete variational LDA method, which has been described in detail elsewhere [36,37]. In formulating the Kohn–Sham equations, we used an exchange and correlation potential of the von Barth–Hedin form [38] with parameters taken from Moruzzi et al. [39]. The atomic basis functions representing the valence electron orbitals were $2s$–$2p$ for C, $3s$–$3p$ for Si, and $1s$ for H, and the rest of the core orbitals were treated as frozen. Using 600 sample points per atom in the numerical integration, we achieved sufficient convergence for both the electronic spectrum and the binding energy. A self-consistent charge model density was used to fit the electron density. We only optimized the height of C_{60} above the Si surface to reduce computational efforts. Finally, the LDOS distribution on the upper hemisphere was calculated with a 16 a.u. radius (R_{t-s}) from the center of the C_{60}.

Our simulated STM images show that the positive bias images depend strongly on the orientation of the C_{60} on the substrate, and less on the adsorption sites. As examples, Figs. 4.11a and b show the calculated images with the sample biased at +2.5 V ($V_s = 2.5$ V) and the C_{60} molecule with one of its 5–6 bonds adsorbed on A and B sites, respectively. Figures 4.11e and f are the calculated images with the same sample bias ($V_s = 2.5$ V) but the C_{60}

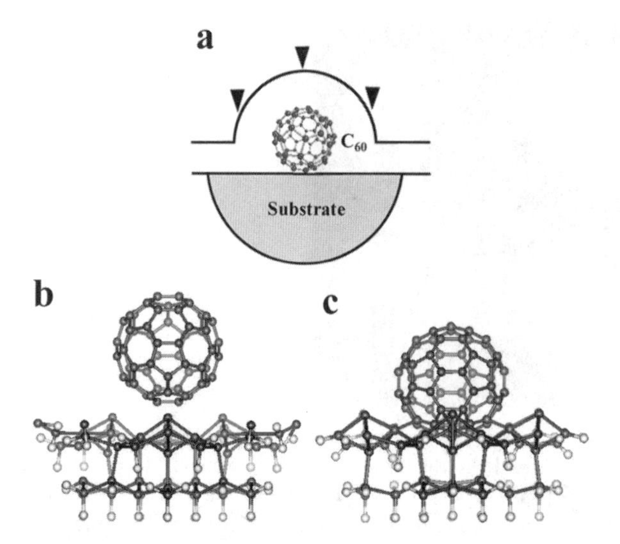

Fig. 4.10. (**a**) Schematic view of the curved surface on which the tip moves around. (**b**) Cluster model of $Si_{57}H_{42}$ for C_{60} adsorbed on site A of an Si(111)–(7×7) surface. (**c**) Cluster model of $Si_{67}H_{45}$ for C_{60} adsorbed on site B

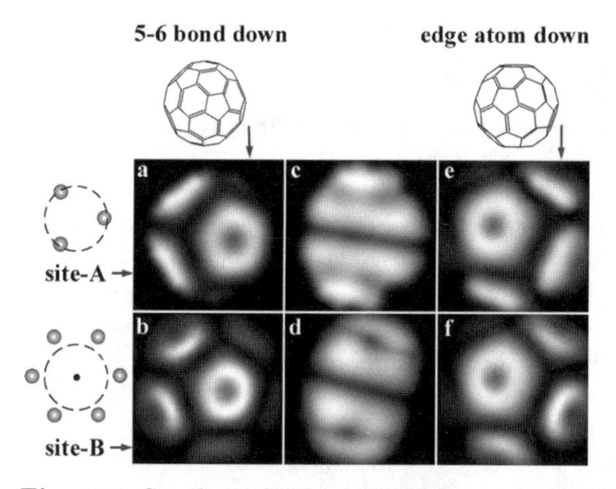

Fig. 4.11. Simulated STM images of C_{60} on an Si(111)–(7 × 7) surface. (**a**) V_s = 2.5 V, 5–6 bond adsorption on A-site. (**b**) V_s = 2.5 V, 5–6 bond adsorption on B-site. (**c**) V_s = −1.8 V, 5–6 bond adsorption on A-site. (**d**) V_s = −1.8 V, edge-atom adsorption on B-site. (**e**) V_s = 2.5 V, edge-atom adsorption on A-site. (**f**) V_s = 2.5 V, edge-atom adsorption on B-site. R_{t-s} = 16 a.u. for all cases

molecules have the edge atom adsorbed on A and B sites, respectively. On the other hand, the simulated negative bias images are only weakly dependent on either the adsorption site or the orientation, and all display four bright stripes. Two calculated images with the sample biased at −1.8 V on sites A

(5–6 bond adsorption) and B (edge-atom adsorption) are given in Figs. 4.11c and d, respectively.

Comparing our simulations with experiment, we find that the experimetal image of C_{60} on the A-site (Fig. 4.9a2) best matches with Fig. 4.11a (a bright pentagon ring plus two bright curved strokes to its left), while the image of C_{60} on the B-site (Fig. 4.9b2) matches well with Fig. 4.11f (one bright pentagon ring plus three bright curved strokes to its right). We thus conclude that C_{60} adsorbs on site A (or A′) with one of its 5–6 bonds facing towards the surface, and on site B with one of its edge atoms facing towards the Si substrate.

The ability to identify the orientation of C_{60} molecules on specific adsorption sites allows us to calculate the local electronic structures and theoretical STS results of oriented C_{60} at those sites, and thus leads to a deeper understanding of the C_{60}–Si interaction. Figure 4.12a shows the energy states of a C_{60} molecule adsorbed at an A-site with one of its 5–6 bonds facing downwards. We noted that the highest occupied energies of a $C_{60}Si_{63}H_{56}$ cluster and a free C_{60} molecule are −7.67 eV and −9.64 eV, respectively. This means that electrons transfer from the Si substrate to C_{60}. From the energy dia-

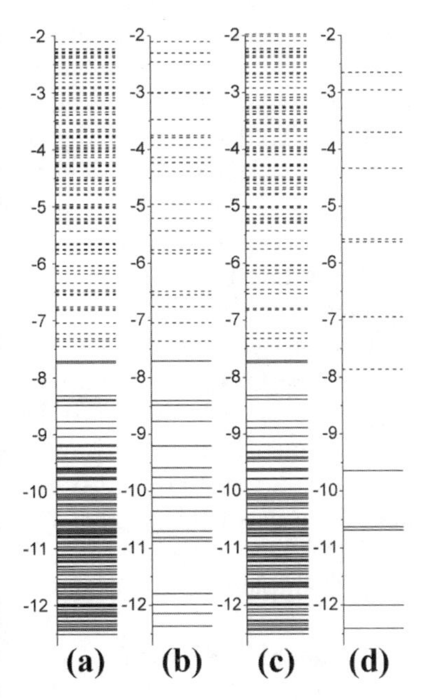

Fig. 4.12. (**a**) Energy levels of a C_{60} molecule adsorbed at an A-site with one of its 5–6 bonds facing downwards. (**b**) Energy levels for C_{60} molecule providing a major Mulliken population. (**c**) Energy levels for Si atoms providing a major Mulliken population. (**d**) Energy levels of a free C_{60} molecule. For all cases, *solid* and *broken lines* denote occupied and unoccupied molecular orbitals, respectively, and units are eV

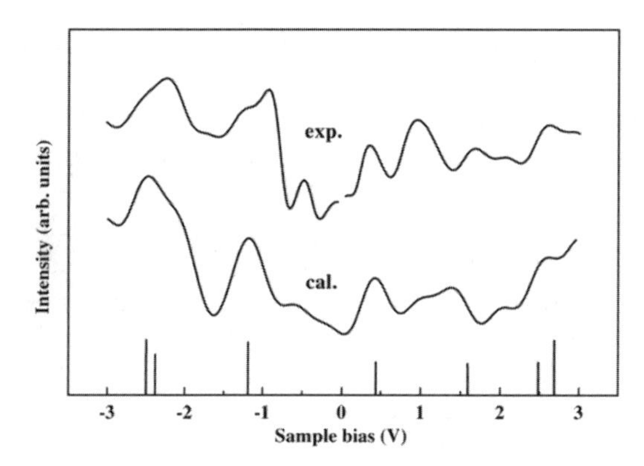

Fig. 4.13. Experimental STS and calculated LDOS of C_{60} adsorbed at site B on an Si(111)–(7 × 7) surface. The *vertical lines* at the bottom are the calculated energy levels of a free C_{60} molecule

gram, we also find that the valence and conduction bands are wide. This may be caused by the hybridization of C and Si atoms.

The energy diagram for C_{60} adsorbed on the B-site is as complicated as that for the A-site shown in Fig. 4.12a. We have calculated the LDOS of C_{60} adsorbed on the B-site and the result is shown in Fig. 4.13 together with the experimental STS. We find that the theoretical result has reproduced the main features of the measured STS. We also plotted the energy levels of a free C_{60} molecule calculated by the discrete-variational local density approximation (DV-LDA) method in Fig. 4.13. For every energy level near the Fermi energy of the free C_{60}, we find a corresponding peak from the calculated LDOS, as well as from the experimental STS. This suggests that the basic characteristics of free C_{60} molecules have not been changed after they adsorb on the Si(111)–(7 × 7) surface. It also proves that it is feasible to determine the HOMO′ and LUMO′ of C_{60} adsorbed on the Si surface. In the calculated LDOS, the mixing states are also found and are similar to the experimental STS result. The calculation gives the ionicity of the entire C_{60} as about −0.5, indicating charge transfers from the Si surface to the adsorbed C_{60} molecule, as confirmed by our STS results. Since it is believed that the bonding between C_{60} and Si is of covalent type [24,40], so the charge transfers from the Si surface to the C_{60} molecule should be implemented through a polar bonding mechanism.

It is interesting to note that although site B is the one with the strongest interaction between C_{60} and the Si surface, it is not the most commonly chosen adsorption site. Based on the above discussion, we suggest here a possible explanation for this. At site B, the two bonding sp^3 hybridized C atoms are separated by a larger distance on the C_{60} molecule, so the formation of these two bonds will need some extra energy to cause the originally nearest two sp^3 hybridization orbitals, initiated by breaking a double bond, to relax onto two separated C atoms in the C_{60} molecule. This means that, even though the interaction between C_{60} and the Si surface is strong at the B-site, it is difficult for a C_{60} molecule to become chemically adsorbed there.

4.3.2 C_{60} Molecules on an Si(100)–(2×1) Surface

Hashizume et al. [32] first studied C_{60} adsorption on the Si(100)–(2 × 1) surface. It was observed that almost all the C_{60} molecules are absorbed in the trough between two dimer rows without any preference to the step edges or nucleation into islands at defect sites. Along the troughs, C_{60} molecules are distributed randomly without forming islands. Indeed the interaction between C_{60} and Si substrate is quite strong since C_{60} molecules stay almost at the place they first land with little mobility.

The STS of individual C_{60} molecules on two different adsorption sites of the Si(100) surface was investigated by Yao et al. [33,34]. In Fig. 4.14 spectrum A is the STS data measured on the molecule located at the top of a single dimer (site A in Fig. 4.15). The occupied state (HOMO) peak and the unoccupied state (LUMO) peak are located at −0.7 eV and −2.3 eV, respectively. The HOMO–LUMO energy gap is thus about 1.8 eV, which is about the same as that for free C_{60} molecules (1.6–1.9 eV). The wide peak at −2.3 eV (labeled I) corresponds to the two closely spaced h_g and g_g levels, which are separated from the HOMO by about 1.6 eV. The t_{1g} peak is found to be about 0.9 eV (labeled IV) above the LUMO level. STS data for C_{60} molecules on site A shows that they have roughly the same energy levels as free C_{60} molecules. This result implies that the breaking of one C–C bond does not have a major influence on the electronic structure of a type-A molecule.

Spectrum B in Fig. 4.14 is the STS data measured for another type of C_{60} molecule which is usually located at a missing dimer defect (denoted as

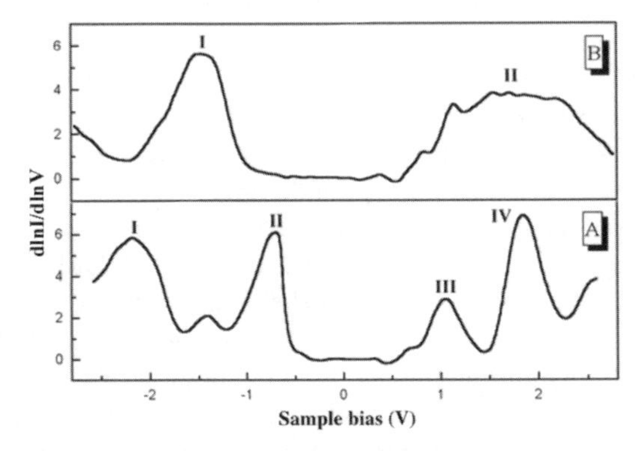

Fig. 4.14. STS measurements on C_{60} molecules located at the top of a single dimer and a missing dimer defect of an Si(100)–(2 × 1) surface, respectively [33]

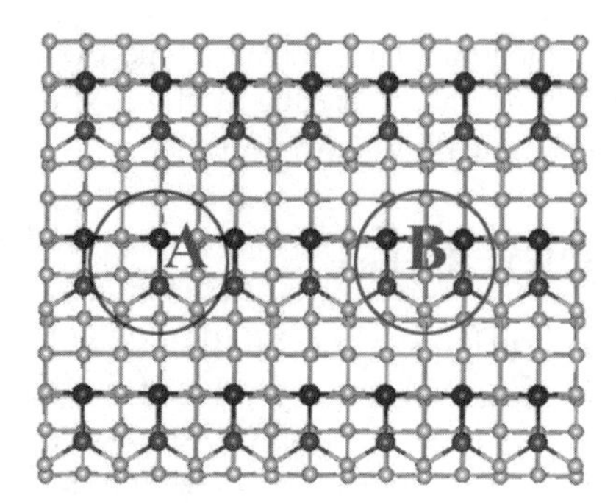

Fig. 4.15. Geometrical model for C_{60} adsorbed on an Si(100)–(2 × 1) surface. *Open circles* show possible adsorption sites

a B molecule). We find a strongly occupied state located at −1.5 eV (labeled I) below E_F and a wide band structure above E_F (labeled II). The gap is about 2.2 eV, which is close to the bandgap of SiC [41,42]. Local atomic rearrangement of the Si(100) surface, which includes a missing dimer defect accompanied by two C-shaped defects in neighboring dimer rows, can be found in the vicinity of every type B molecule. STS results indicate that the interaction between a type B molecule and the surface is much stronger than that for an A molecule. The result also suggests that B molecules form covalent bonds not only with second layer Si atoms, but also with the surrounding Si atoms in the first layer.

Hashizume et al. [32] observed three or four bright strips for each C_{60} in their high resolution negative bias STM images. The STM images reported by Yao et al. [33,34] provided a few non-strip patterns. Their results indicated that the intramolecular structure of type A molecules is very different from that of type B molecules and supported the STS conclusion that the local electronic structure of the A-site differs from that of the B-site. However, Yao et al. did not discuss C_{60} molecular orientations at different sites.

Several investigations have been done to try to understand the origin of these strip-like patterns. Kawazoe et al. [43] calculated the electronic structure of the C_{60} molecule on the Si(100)–(2 × 1) surface by means of a first-principles calculation. They took one layer of C_{60} molecules, and the Si(100) substrate was treated as a positive charge background. According to their calculation, the electron charge density of distribution for each level can be mapped out and the charge density distribution of the 180th level (HOMO) and the 181st level (LUMO) exhibit strips similar to those in the STM images. Yamaguchi [44] calculated the electronic structure of a model cluster of C_{60} molecules on an Si(100)–(2 × 1) surface by the DV-X_α method. Yajima et al. [45] pointed out that it is the local density of states which should be compared with STM images, and that it is not sufficient to calculate the electron density contour map of some particular energy states. Although their results reproduced the strip-like patterns in a way similar to the experimen-

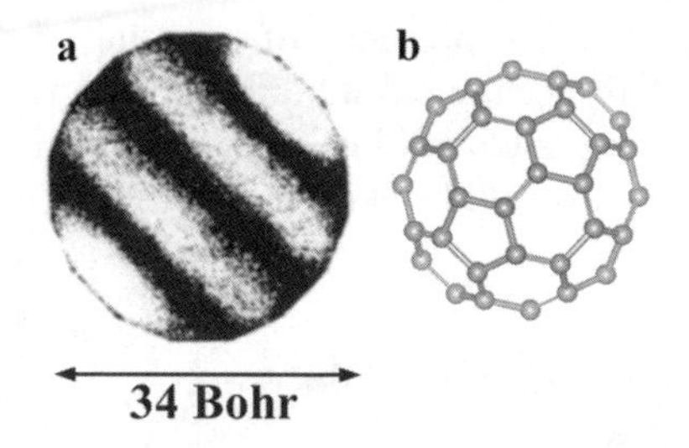

Fig. 4.16. (**a**) Simulated STM image of the C_{60} molecule on an Si(100)–(2 × 1) surface (sample bias -2.0 eV, tip height 12 bohr) [45] and (**b**) the corresponding C_{60} orientation

tal results (see Fig. 4.16), it is still difficult to identify molecular orientations since the negative bias images are not dependent on molecular orientation and adsorption site, as shown above. Further experiments are needed to obtain high resolution positive bias images of the C_{60}/Si(100) system if we are to understand the local C_{60}–Si interaction mechanism.

4.4 Electronic Transport Properties of a Single Fullerene Molecule

Until recently, the electronic properties of individual molecules had not been measured. The main difficulty is the question of how to connect individual molecules to electrodes. The scanning probe microscope has dramatically changed this situation by introducing a molecule into tunnel junctions in which the tip and supporting surface are two electrodes. The STM tunnel junction with a molecule at the interface has the advantage that single molecules can be observed on an atomically flat electrode (surface) before or after the electrical measurement.

Molecules are typically characterized by their HOMO and LUMO states separated by an energy gap. The discrete levels of the HOMO and LUMO states are broadened and shifted through their interaction with the surface when the molecule is adsorbed on a substrate. When a voltage is applied across a molecule by two electrodes, electrons tunnel directly through the LUMO states and are therefore in resonance with these states if the bias voltage is larger than the energy gap. If the applied bias voltage is not high enough to allow electrons to tunnel directly through the molecular levels, electrons can still be transported by means of the tails of the broadened molecular levels. This transport mechanism, known as virtual resonance tunneling (VRT), can be used to produce an electronic amplification phenomenon in single molecules.

Recently, electronic properties of individual C_{60} molecules have been investigated using STM [7,46,47]. The resistance of a C_{60} molecule was determined by defining the 'tip contact' with the molecule and carrying out tunneling spectroscopy on an isolated C_{60} in the presence of charging effects. Single-molecule amplification by electromechanical modulation of virtual resonance quantum mechanical tunneling was also investigated, and it was shown that the STM tip can make, visualize and operate such a device rather than just function as an electrode.

4.4.1 Electrical Resistance of a Single C_{60} Molecule Determined by STM

When the STM tip is very close to a conducting surface, the electrical conductance (resistance) of the metal–vacuum–metal junction is given by the slope of the I–V characteristic at low applied voltages where a linear relationship is observed. Typically, the voltage (V_t) is less than 100 mV [48]. If a molecule is introduced into such a junction, the I–V characteristic will certainly be changed, and in principle the electrical resistance of the molecule can be determined accordingly. However, some basic concepts regarding the electrical transport of a single molecule between two metallic electrodes have to be addressed.

Büttiker and Landauer [49] have developed a concept of 'electronic transparency' to characterize the ability of a molecule to let electrons flow through it. High transparency corresponds to a low resistance of the metal–molecule–metal junction. In contrast to macroscopic conductance determined by electron–phonon scattering, the transport of electrons through a molecule is determined by the quantum mechanical probability of one electron moving from one electrode to the other. As discussed above, if a molecule's HOMO and LUMO states do not lie in the range of the Fermi level, tunneling occurs by a process known as VRT when the bias voltage is low enough. In this case, transparency is determined by the efficiency of coupling between the molecule and its metallic substrate or the non-zero weight of the tails of the broadened molecular levels at the electrode's Fermi level.

Consequently, the issue of the electronic conductance or transparency of a single molecule requires us to understand the molecule–substrate interaction and its influence on the broadening and shifting of molecular levels. In particular, when the STM tip approaches the molecule and the tip–molecule junction becomes smaller and smaller, molecular orbitals can be modified by mechanical deformation of the molecular conformation or shape, thereby also affecting the electronic coupling of molecule and electrodes. It is expected that such deformation and modification will significantly change the slope of the I–V characteristics, so that the meaning of electronic and mechanical contacts with an individual molecule should be defined in such a way as to get a reasonable value of single-molecule conductance (resistance) from the experimental results.

Joachim et al. [46] did the first study of electrical contact with an individual C_{60} molecule. The experiment was performed using an STM at 300 K in ultrahigh vacuum. Submonolayer C_{60} molecules were deposited by thermal

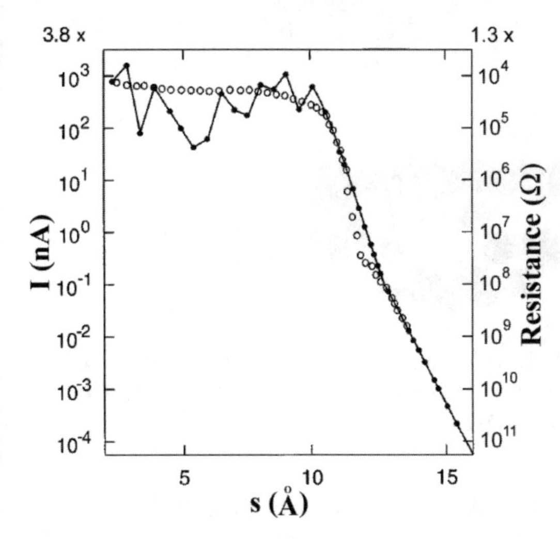

Fig. 4.17. Variation of the tunneling current intensity $I(s)$ through a C_{60} molecule and corresponding resistance as a function of the distance s from tip apex to surface. *Open circles* indicate experimental values. The *solid line* indicates calculated values with full deformation of the C_{60} upon tip approach. In both cases, a bias voltage of 50 mV was chosen [46]

evaporation onto an Au(100) surface. The C_{60} HUMO–LUMO gap is not too wide in its adsorbed state on Au (about 1 eV) [50], so a reasonable current intensity can be measured at low bias voltage within the VRT regime. The tip was positioned above a C_{60} and the electrical current I was studied as a function of tip displacement s towards the molecule [$I(s)$ characteristic]. This was achieved by superimposing a series of offsets Δs and recording individual I–V characteristics over a range of resistances varying from $R \approx 5\,\mathrm{M\Omega}$ to $R \approx 5\,000\,\mathrm{M\Omega}$. It was confirmed that, to within the bias voltages of $\pm 200\,\mathrm{mV}$, I–V characteristics recorded at fixed s are linear for a series of tip altitudes. This linearity allows us to define the tunneling resistance by $I(s)$ measurements at constant $V_t = 50\,\mathrm{mV}$. Figure 4.17 shows a representative $I(s)$ curve averaged over some 20 measurements at a bias voltage of 50 mV. All the $I(s)$ curves display a linear $\log I$ variation as the tip is ramped forward, followed by a large and reversible positive deviation at smaller gap values.

Joachim et al. [46] also calculated the theoretical $I(s)$ curve using the elastic scattering quantum chemistry technique configured for STM (STM-ESQC) [51]. The full atomic structure and valence orbitals of the tip, the C_{60} molecule, and the Au(110) substrate were considered in these calculations. The C_{60} deformation was included in STM-ESQC using the MM2 routine with a standard sp^2 carbon parametrization [52], and the C_{60} conformation was optimized for each s. The current I through the molecule was then calculated for each conformation.

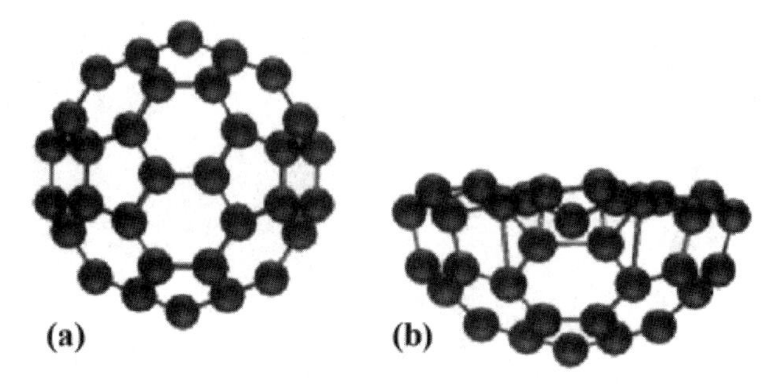

Fig. 4.18. Optimized structure of the C_{60} molecule in the tunneling junction for a tip apex to surface distance of (**a**) $s = 14\,\text{Å}$ and (**b**) $s = 7.35\,\text{Å}$. The W tip apex was considered to be rigid during approach [46]

The deformation of C_{60} by the W tip apex is presented in Fig. 4.18, as optimized during STM-ESQC calculations. The calculated $I(s)$ curve (solid line in Fig. 4.17) agrees very well with the experimental results when junction distances are between 16 and 14 Å. As s decreases down to 13 Å, a small deformation of the C_{60} cage occurs due to van der Waals tip attraction and the deformation leads to a deviation from the linearity of $\log I(s)$. In the range 16–13 Å, the junction is in a tunneling regime. According to the optimized energy of C_{60} in the junction, mechanical contact is established at $s \sim 13.2\,\text{Å}$, corresponding to zero force on the C_{60} in the junction $[\mathrm{d}(E(C_{60})/\mathrm{d}s = 0]$. At this point the C_{60} is slightly attracted by the tip. At $s = 12.3\,\text{Å}$, the van der Waals expansion is compensated by tip compression and the C_{60} recovers approximately its original shape. If this point is defined as the point of electrical contact, the resistance of the total junction with C_{60} is $54.807 \pm 13\,\mathrm{M\Omega}$ as determined from experimental measurement (see Fig. 4.17). When $s < 12.0\,\text{Å}$, the cage undergoes compression and $R(s)$ decreases rapidly towards the quantum resistance limit of $R_0 = h/2e^2$. At this limit the molecule is almost totally transparent to the tunneling electrons.

Is the resistance of $54.807\,\mathrm{M\Omega}$ the intrinsic electrical resistance of an individual C_{60} molecule? At low bias voltage, all the I–V characteristics remain linear within the HOMO–LUMO gap and the tunneling current through C_{60} is dependent on the symmetry and spatial extension of its molecular orbits [51]. In this sense, the resistance may be considered to contain an intrinsic characteristic of the molecule. Nevertheless, the current flows because the C_{60} molecular orbitals are broadened and shifted by their electronic coupling with the tip and surface. Joachim's observation of a pressure dependence of the electronic transparence of an individual C_{60} is a step toward a realistic characterization of the electrical properties of a single molecule in a nanoscopic environment. Moreover, this study indicates key differences in the physics of C_{60} confined in such a situation compared to bulk solid C_{60} [53,54].

4.4.2 Resonant Tunneling of Single C_{60} Molecules in a Double Barrier Tunneling Junction (DBTJ)

Metal clusters with a size of a few nanometers have been intensively studied [55–66]. When the metal cluster is placed as the center electrode of a DBTJ, the current–voltage (I–V) curves of the structure show Coulomb blockade and Coulomb staircase behavior due to the discreteness of the charge and the significant charging energy E_C of the cluster as compared to the thermal energy k_BT [55–63,66]. When the size of the metal cluster is small enough, the quantum size effect further quantizes the energy levels which can be observed in the I–V characteristic of a DBTJ as well [64–66]. The interplay between the single-electron tunneling (SET) effect and the quantum size effect can be experimentally observed most clearly when the charging energy E_C of the cluster by a single electron is comparable to the electronic energy level separation E_L, and both energy scales are larger than k_BT. Both effects are relevant to the development of nanoscale electronic devices, such as single-electron (and single-molecule) transistors, and have thus attracted considerable experimental and theoretical attention [67–69].

An isolated C_{60} molecule provides an ideal model system for this research. It is unique due to its size (~ 8 Å in diameter), much smaller than the nanoparticles investigated so far, its spherical shape, and its rich electronic spectrum, which is affected by charging. Moreover, this system enables a simultaneous study of the interplay between two regimes, one where the level spacing $\Delta E_L > E_C$ (due to molecular orbital spacing), and the other where $\Delta E_L < E_C$ (due to level splitting).

Porath et al. studied the tunneling spectroscopy of isolated C_{60} molecules in the presence of charging effects using a cryogenic STM [7]. They realized a DBTJ configuration in which a C_{60} molecule is coupled via two tunnel junctions to the gold substrate and the tip of the STM (see Fig. 4.19a), and a thin insulating polymethyl methacrylate (PMMA) layer (less than 40 Å thick) was used as the barrier between C_{60} and gold substrate. The C_{60} molecules were studied spectroscopically by taking tunneling I–V characteristics. Figure 4.19b plots four I–V characteristics for a single molecule at 4.2 K with different STM settings (shifted for clarity). The main features in the I–V curves are as follows. Firstly, there is a nonvanishing gap in the curves around zero bias that oscillates periodically between minimum and maximum widths as we monotonically change I_s for fixed V_s (and vice versa). This oscillation is due to the periodic variation of 'fractional charge' (Q_0), consistent with the 'orthodox' theory [70,71]. The fact that the minimum gap is nonzero indicates an additional contribution related to the molecular level spacing besides the 'classical' Coulomb blockade. Moreover, in contrast to pure SET characteristics, the curves become highly asymmetric and exhibit steps with variable widths and heights.

Figure 4.20a plots three experimental I–V characteristics (thick curves) taken with different junction parameters such that $C_1 < C_2$, $C_1 \approx C_2$ and

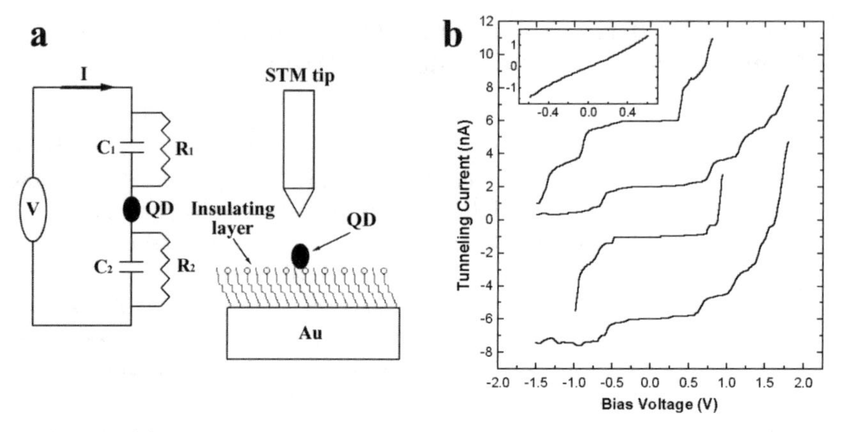

Fig. 4.19. (**a**) Schematic and equivalent circuit of the experimental setup: a C_{60} molecule [quantum dot (QD)] coupled via two tunnel junctions to the STM tip and gold substrate. (**b**) Tunneling I–V characteristics at 4.2 K taken for a C_{60} molecule in a DBTJ configuration, with different STM settings. *Inset*: an I–V trace taken near a C_{60} molecule showing Ohmic behavior [7]

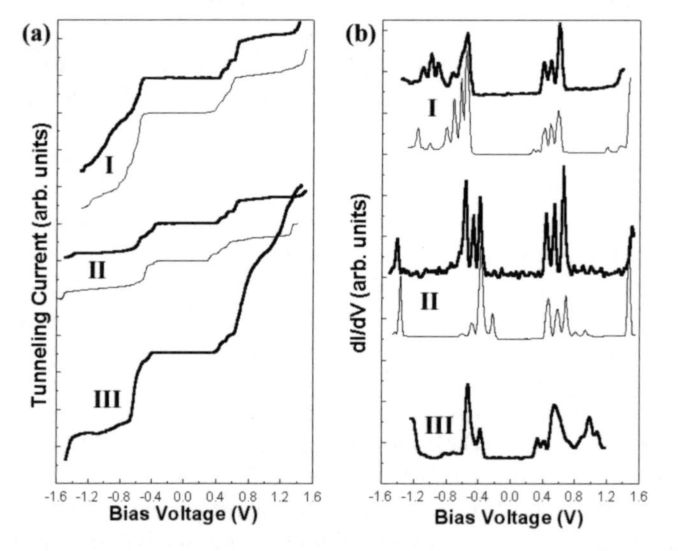

Fig. 4.20. Tunneling I–V characteristics and their $\mathrm{d}I/\mathrm{d}V$ traces at 4.2 K [*thick curves* in (**a**) and (**b**)] and corresponding theoretical fitting [*thin curves* in (**a**) and (**b**)] for C_{60} in a DBTJ. Curves group I: $C_1 < C_2$. Curves group II: $C_1 \approx C_2$. Curves group III: $C_1 > C_2$ [7]

$C_1 > C_2$, respectively, where C_1 is the capacitance of the molecule–electrode junction and C_2 the capacitance of the tip–molecule junction. According to the 'orthodox' theory [70,71], the step width from the SET effect is proportional to the reciprocal of the larger of C_1 and C_2. From Fig. 4.20, we find that these curves contain additional steps beyond those related to the SET effect, which reflect the discrete electronic structure of the C_{60} molecule. If the data is plotted in the dI/dV traces, shown in Fig. 4.20b, the structure is exhibited in a more pronounced way in which each peak corresponds to resonant tunneling through a discrete level.

Porath et al. also did theoretical calculations to explain the experimental spectrum. They used 'orthodox' theory [70] but modified the formalism to include the electronic spectrum of C_{60}. Information on both the junction parameters and the electronic molecular spectrum was obtained from the calculation. For an unperturbed, neutral C_{60} molecule, the spherical $L = 5$ levels are split into three groups: fivefold degenerate h_{1u} levels (HOMO), threefold degenerate t_{1u} (LUMO), and threefold degenerate t_{2u} (LUMO+2). The LUMO+1 (t_{1g}, threefold degenerate) originates from the $L = 6$ levels. These degeneracies can be lifted by two different means that break icosahedral symmetry: the electric field between the tip and the substrate, and the Jahn–Teller (JT) effect [72,73], which is prominent in the ionized and excited states of the molecule. Accordingly, they fitted all the curves using three groups of levels related to the HOMO, LUMO, and the LUMO+1 orbitals. The results are also represented in Fig. 4.20. One can observe a full splitting of the five HOMO levels and the three LUMO levels as well as a partially lifted degeneracy in these levels. The fittings were performed by taking the HOMO levels in the range -0.8 to -0.6 eV, the LUMO in the range 0–0.15 eV, and the LUMO+1 onset at ~ 0.6 eV. It should be noted that the observed HOMO–LUMO level spacing is ~ 0.7 eV, much smaller than the level spacing in the free neutral molecule. This result is reasonable because they are actually measuring the gap between the LUMO (C_{60}^-) and the HOMO (C_{60}^+), and because the threshold for tunneling in the DBTJ is determined by the energy of the final, ionized state of the molecule. The typical level splitting they obtained, ~ 0.05 eV, is somewhat smaller than predictions for the JT effect [72,73].

The effect of the junction capacitances on the shapes of the I–V curves is also an interesting question. The calculated results indicate that both the charging energy and the level separation exceed the thermal energy k_BT even at room temperature, so the SET effect and the interplay with the molecular levels can be clearly observed [74]. When $C_1 < C_2$, tunneling through the DBTJ commences at junction 1, for the level configuration at hand, where electrons tunnel into the LUMO for positive bias polarity, and off the HOMO for negative bias. This is a major source for the pronounced asymmetry observed in the curves. The above interpretation is supported by the fact that in the opposite case, when $C_1 > C_2$, the I–V curves obtained are nearly an inversion (with respect to $V = 0$) of those in Fig. 4.20.

In summary, the discrete tunneling spectra of isolated C_{60} fullerenes in the DBTJ configuration manifest the interplay between charging and quantum size effects simultaneously in two regimes: $\Delta E_L > E_C$ and $\Delta E_L < E_C$.

The degenerate HOMO, LUMO, and possibly the LUMO+1 levels of the unperturbed molecule were split and fully resolved spectroscopically. The degree of splitting is somewhat smaller than predictions for the JT effect. Local fields may contribute to this degeneracy-lifting as well. Theoretical curves, calculated using the 'orthodox' model for SET, modified to account for the discrete levels spectrum of C_{60}, agree well with the experimental data. The work of Porath et al. indicates that, in spite of the small size of the molecule, one can associate with it an effective capacitance similar to that of metallic particles of comparable size. A quantum-mechanical approach is probably more suitable for treating the capacitance in this molecular size regime.

4.4.3 Molecular Device Using a Single C_{60} Molecule as the Functional Unit

Design and fabrication of molecular devices based on their amazing functional quantum properties using STM has been an inspiring field recently [75,76]. Cage-structured C_{60} molecules have many interesting properties and have shown their amazing potential in applications to molecular devices.

Joachim and Gimzeweski have proposed and demonstrated a single-molecule amplifier [77]. It is based on an extension of their measurements discussed in Sect. 4.4.1 as regards the electromechanical modification of VRT for a single C_{60} molecule contacted by two electrodes. The principle of the device can be described in more detail as follows. When the tip apex approaches the top of the C_{60} cage, electrical contact between the molecule and the electrode is defined as the point when the tunneling current intensity is maximum through C_{60} while maintaining the condition of minimum deformation of the C_{60} cage. Such a condition is realized by selecting V_{g} (the gate voltage of the amplifier) in such a way that the HOMO–LUMO gap is almost that of an unperturbed C_{60} molecule. When the tip is further lowered by 0.1 nm toward the molecule, cage deformation leads to an enhancement of current intensity by two orders of magnitude. Within this 0.1 nm range, the I–V characteristic of the device remains linear if V_{p} (the voltage through the polarization resistance) is kept below 200 mV. By optimizing the load and polarization resistances (R_{L} and R_{P}) according to the C_{60} resistance, under and out of compression, a gain of five has been obtained experimentally in such a molecular amplifier. This work demonstrates a new mechanical means for controlling quantum mechanical processes on the subnanometer scale. Moreover, such a device has the amazingly tiny operation energy of one attojoule (10^{-18} J) and this value is four orders of magnitude smaller than the best known solid state switch [78].

The Esaki tunneling diode [79], which has a wide variety of applications in the semiconductor industry, such as fast switches, oscillators, and frequency-locking circuits, has also been exploited at nanometer scales. The negative differential resistance (NDR) characterized by the phenomenon of decreasing current with increasing voltage in the I–V curves is the essential property of the Esaki diode. It has been shown experimentally [80–85] and theoretically [85,86] that NDR could arise in the tunneling structures with an STM tip and a localized surface site when the LDOS of both the tip and the surface site have sufficiently narrow features. The sharp LDOS features of the surface site have been realized by, e.g., doping B atoms on the Si(111) surface, [80–82], depositing Ag atoms on a semiconductor surface [83], using rebonded B-type steps on the Si(001)–2×1 surface [84], or synthesizing a self-assembled monolayer of small organic molecules on a gold substrate [85]. The narrow LDOS features of the tip are usually realized by 'sharpening' the tip apex, which means attaching a foreign adatom or small cluster. We note that the structure of such a 'sharpened' tip and the associated LDOS was uncontrollable and unstable to a certain extent in previous experiments [80,81,83]. Thus NDR is not always observable over a particular type of site, but depends on the tip status which may vary even within a single experiment [80,81].

We propose here an NDR molecular device involving a Pt–Ir tip with a C_{60} molecule on its apex and another C_{60} molecule physically adsorbed on a hexanethiol SAM. Figure 4.21a shows the schematic diagram of the NDR device. Previous experiments showed that a C_{60} molecule could easily be attached to the tip when a Pt–Ir tip was brought near the sample surface and rastered through C_{60} islands [87]. Because of the enhanced metal–C_{60} interactions [87,88], the attached C_{60} molecule can be stable at the apex of the tip. It is well known that a free C_{60} molecule has I_{h} symmetry and narrow-featured electronic structure. The interaction between the adsorbed C_{60} molecule and the Pt–Ir tip is not strong enough to destroy the molecular symmetry [88,89], so the narrow features of the electronic states of C_{60} are preserved. As for the C_{60} molecules adsorbed on the hexanethiols, they are weakly bound to the substrate by van der Waals forces. Consequently, the sharp features of the LDOS of the C_{60} on the hexanethiol surface can also be preserved. We therefore suggest that the NDR phenomena can be observed due to the narrow structure in the LDOS of the C_{60} molecules on both electrodes.

In order to confirm that NDR could be realized from the tunneling between two C_{60} molecules, we performed a theoretical simulation for the tunneling structure shown in Fig. 4.21a. As a first order approximation, the tunneling current can be expressed as [90]

$$I(V) \propto \int_{E_{\mathrm{F}}}^{E_{\mathrm{F}}+eV} \rho_{\mathrm{T}}(E-V)\rho_{\mathrm{S}}(E)\exp\left[-2s\sqrt{2(W-E)+V}\right]\mathrm{d}E\ , \quad (4.3)$$

where $\rho_{\mathrm{T}}(E)$ and $\rho_{\mathrm{S}}(E)$ are the LDOS associated with the tip and sample respectively, W is the height of the tunneling barrier, and s is the tip–sample distance.

The tunneling structure in Fig. 4.21a consists of two junctions: junction 1 is between C_{60} on the tip and C_{60} on the hexanethiols, and junction 2

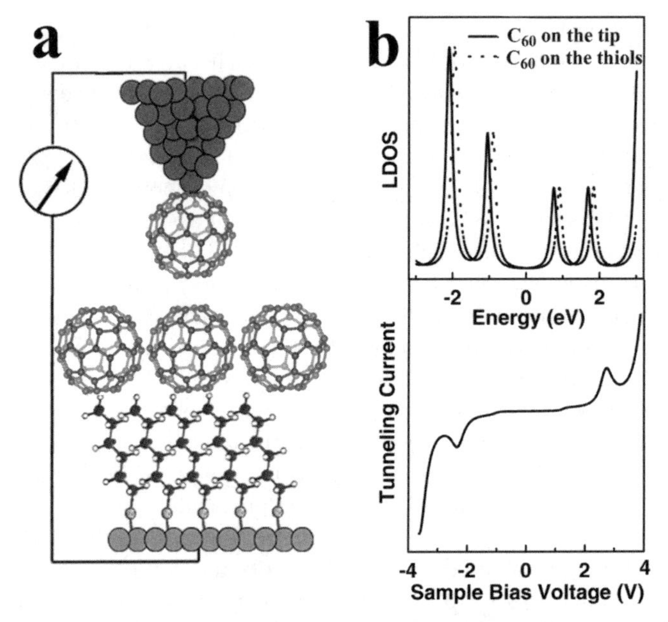

Fig. 4.21. (**a**) Schematic diagram of the NDR device involving one C_{60} molecule physically adsorbed on the SAM and another adsorbed on an STM tip. (**b**) The simulated LDOS (*upper part*) of C_{60} on the tip and C_{60} on the SAM, and corresponding calculated I–V curve (*lower part*)

between C_{60} on the hexanethiols and the Au(111) substrate. The voltage drop across junction 1 is

$$V_1 = \frac{VC_2}{C_1 + C_2}, \tag{4.4}$$

where V is the applied bias voltage, and C_1 and C_2 are the electric capacitances of junction 1 and junction 2, respectively. Considering that the separation between the centers of C_{60} on the tip and C_{60} on the hexanethiols is assumed to be 10.5 Å, the thickness of the hexanethiol SAM is about 10.2 Å [91], the dielectric constant for SAM on Au(111) is taken as 2.6 [92], and V_1/V is estimated to be 0.72. In the simulation, the energy levels of a free C_{60} molecule were calculated by the DV-LDA method. Then the LDOS of the C_{60} molecules on both electrodes were derived from the Lorentz broadening of these discrete energy levels [93]. The Lorentzian width depends on coupling with the environment, as well as on the temperature. Figure 4.21b shows the LDOS of the C_{60} molecules obtained by assuming a constant Lorentzian width (0.18 eV) for all the energy levels. The shift of the LDOS of the C_{60} on the tip is due to charge transfer from the tip to the C_{60} molecule [87,88]. In the present work, we assume a 0.15 eV shift of the Fermi level toward the unoccupied states. The simulated I–V curve clearly shows the NDR effect at both positive and negative bias.

From the simulation, we can conclude that an NDR molecular device can be realized by the tunneling structure involving a C_{60} adsorbed on a metal tip and another C_{60} weakly bound to the hexanethiol SAM. This is attributed to the narrow-featured LDOS near the Fermi level of both the C_{60} molecules. Compared with the previous atomic scale NDR device, the advantage of the proposed C_{60} molecular NDR device is that it has a controllable and relatively stable tip structure and therefore stable and reproducible NDR is expected in such a device.

4.5 Conclusion

We have reviewed progress in the study of single fullerene molecules using STS. From the STS data we can determine site- and orientation-dependent electronic structures of single molecules adsorbed on a surface. In principle, the method described in this section can also be applied to other kinds of molecule–substrate system. However, we should interpret STS data with caution since the electronic structure of the STM tip is to some extent uncontrollable and so may strongly modify the measured spectrum. Elaborate spectroscopic experiments should therefore be combined with high resolution STM images and theoretical modeling in order to eliminate the tip effect and thus extract intrinsic information about the adsorbed single molecule. Moreover, experiments on individual molecules using STM have permitted deeper insight into the quantum electronics of molecular systems such as the interplay between the single-electron tunneling effect and splitting of degenerate molecular levels. At the level of an individual molecule, unique information has been obtained about conformational and mechanical properties and their relations to quantum-electronical behavior. All this has extended our knowledge of single molecules and may lead to the design and construction of new artificial molecular devices and machines.

References

1. W. Krätschmer, M.S. Dressechaus: *Sciences of Fullerene and Carbon Nanotubes* (Academic Press, New York 1996)
2. D.M. Eigler, C.P. Lutz, W.E. Rudge: Nature **352**, 600 (1991)
3. D. Smith: Science **269**, 371 (1995)
4. Y. Wada, T. Uda, M. Lutwyche, S. Kondo, S. Heike: J. of Appl. Phys. **74**, 7321 (1993)
5. D. Goldhaber-Gordon, M.S. Montemerlo, J.C. Love, G.J. Opiteck, J.C. Ellenbogen: Proc. of the IEEE **85**, 521 (1997)
6. J.K. Gimzewski: Physics World **11** N6, 29 (1998)
7. D. Porath, Y. Levi, M. Tarabiah, O. Millo: Phys. Rev. B **56**, 9829 (1997)
8. G. Binnig, H. Rohrer: Scientific American **253**, N2, 40 (1985)
9. G. Binnig, H. Rohrer, C. Gerber, E. Weibel: Physica B **109/110**, 2075 (1982)
10. G. Binnig, H. Rohrer: Helvetica Physica Acta **55**, 726 (1992)

11. J. Tersoff, D.R. Hamann: Phys. Rev. B **31**, 805 (1985)
12. R.M. Feestra, J.A. Stroscio, J. Tersoff, A.P. Fein: Phys. Rev. Lett. **58**, 1192 (1987)
13. R.J. Hamers, R.M. Tromp, J.E. Demuth: Phys. Rev. Lett. **56**, 1972 (1986)
14. A.L. Vázquez de Parga, O.S. Hernán, R. Miranda, A. Levy Yeyati, N. Mingo, A. Martín-Rodero, F. Flores: Phys. Rev. Lett. **80**, 357 (1998)
15. H.W. Kroto, J.R. Heath, S.C. O'Brien, R.F. Curl, R.E. Smalley: Nature **318**, 162 (1985)
16. R. Taylor, J.P. Hare, A.K. Abdul-Sada, H.W. Kroto: J. of the Chem. Soc., 1423 (1990)
17. S. Saito, A. Oshiyama, Y. Miyamoto: *Proceedings of Computational Physics for Condensed Matter Phenomena: Methodology and Applications, Osaka, Japan, October 21–23, 1991*, ed. by M. Imada, S. Miyashita et al. (Springer, Berlin, Heidelberg, New York 1992)
18. S. Saito: *Cluster Assembled Materials*, Fall 1990 Materials Research Society Proceedings Boston MA, ed. by R.S. Averback, D.L. Nelson and J. Bernholc
19. For a review, see R.C. Haddon: Accounts of Chemical Research **25**, 127 (1992)
20. T. Sakurai, X.D. Wang et al.: Prog. in Surf. Sci. **51**, 263–408 (1996) and references therein
21. Y. Fujikawa, K. Saiki, A. Koma: Phys. Rev. B **56**, 12124 (1997)
22. S. Suto, K. Sakamoto, T. Wakita: Phys. Rev. B **56**, 7439 (1997)
23. P. Moriarty, M.D. Upward, A.W. Dunn, Y.-R. Ma, P.H. Beton: Phys. Rev. B **57**, 362 (1998)
24. K. Sakamoto, M. Harada, D. Kondo, A. Kimura, A. Kakizaki, S. Suto: Phys. Rev. B **58**, 13951 (1998)
25. O. Janzen, W. Mönch: J. of Phys.: Cond. Matt. **11**, L111 (1999)
26. X.D. Wang, T. Hashizume, H. Shinohara, Y. Saito, Y. Nishina, T. Sakurai: Japanese J. of Appl. Phys. **31**, L983 (1992)
27. Y.Z. Li, M. Chander, J.C. Patrin, J.H. Weaver: Phys. Rev. B **45**, 13837 (1992)
28. J.G. Hou, J.L. Yang, H.Q. Wang, Q.X. Li, C.G. Zeng, H. Lin, B. Wang, D.M. Chen, Q.S. Zhu: Phys. Rev. Lett. **83**, 3001 (1999)
29. H.Q. Wang, C.G. Zeng, Q.X. Li, B. Wang, J.L. Yang, J.G. Hou, Q.S. Zhu: Surf. Sci. **442**, L1024 (1999)
30. S. Saito, A. Oshiyama: Phys. Rev. Lett. **66**, 2637 (1991)
31. R.E. Haufler, R.E. Smalley: Phys. Rev. Lett. **66**, 1741 (1991)
32. T. Hashizume, X.D. Wang, Y. Nishina, H. Shinohara, Y. Saito, Y. Kuk, T. Sakurai: Japanese J. of Appl. Phys. **31**, L880 (1993)
33. X. Yao, T.G. Ruskell, R.K. Workman, D. Sarid, D. Chen: Surf. Sci. **366**, L743 (1996)
34. X. Yao, R.K. Workman, C.A. Peterson, D. Chen, D. Sarid: Appl. Phys. A **66**, S107 (1998)
35. D. Chen, D. Sarid: Phys. Rev. B **49**, 7612 (1994)
36. B. Delley, D.E. Ellis: J. of Chem. Phys. **76**, 1949 (1982)
37. B. Delley, D.E. Ellis, A.J. Freeman, E.J. Baerends, D. Post: Phys. Rev. B **27**, 2132 (1983)
38. U. von Barth, L. Hedin: J. of Phys. C **5**, 1629 (1972)
39. V.L. Moruzzi, J.F. Janak, A.R. Williams: *Calculated Electronic Properties of Metals* (Pergamon, New York 1978)
40. A.J. Maxwell, P.A. Brühwiler, D. Arvanitis, J. Hasselström, M.K.-J. Johansson, N. Martensson: Phys. Rev. B **57**, 7312 (1998) and references therein

41. A.V. Hamza, M. Balooch, M. Moalem: Surf. Sci. **317**, L1129 (1994)
42. D. Chen, R. Workman, D. Sarid: Surf. Sci. **344**, 23 (1995)
43. Y. Kawazoe, H. Kamiyama, Y. Maruyama, K. Ohno et al.: Japanese J. of Appl. Phys. **32**, 1433 (1992)
44. T. Yamaguchi: J. of the Physical Society of Japan **62**, 3651 (1993)
45. A. Yajima, M. Tsukada: Surf. Sci. **366**, L715 (1996)
46. C. Joachim, J.K. Gimzeweski., R.R. Schlittler, C. Chavy: Phys. Rev. Lett. **74**, 2102 (1995)
47. C. Joachim, J.K. Gimzeweski: Proc. of the IEEE **86**, 184 (1998)
48. G. Bining, H. Rohrer: IBM J. Res. Dev. **30**, 357 (1986)
49. R. Landauer: Philosophical Magazine **21**, 863 (1970)
50. J.K. Gimzewski, S. Modesti, R.R. Schlittler: Phys. Rev. Lett. **72**, 1036 (1994)
51. C. Chavy, C. Joachim, A. Altibelli: Chem. Phys. Lett. **214**, 569 (1993)
52. N.L. Allinger, R.A. Kok, M.R. Iman: J. of Comp. Chem. **9**, 591 (1988)
53. M. Núnez-Regueiro, P. Monceau, A. Rassat, P. Bernier, A. Zahab: Nature **354**, 289 (1991)
54. F. Moshary, N.H. Chen, I.F. Silvera, C.A. Brown, H.C. Dorn, M.S. de Vries, D.S. Bethune: Phys. Rev. Lett. **69**, 466 (1992)
55. For a review, see H. Grabert, M.H. Devoret: *Single Charge Tunneling*, NATO ASI Series (Plenum, New York 1991)
56. R. Kubo: J. of the Physical Society of Japan **17**, 975 (1962)
57. R. Wilkins, E. Ben-Jacob, R.C. Jaklevic: Phys. Rev. Lett. **63**, 801 (1989)
58. P.J.M. van Bentum, R.T.M. Smokers, H. van Kenpen: Phys. Rev. Lett. **60**, 2543 (1988)
59. C. Schönenberger, H. van Houten, H.C. Donkersloot: Europhys. Lett. **20**, 249 (1992)
60. C. Schönenberger, H. van Houten, H.C. Donkersloot, A.M.T. van der Putten, L.G.J. Fokkink: Physica Scripta **T45**, 289 (1992)
61. C. Schönenberger, H. van Houten, J.M. Kerkhof, H.C. Donkersloot: Appl. Surf. Sci. **67**, 222 (1993)
62. D. Anselmetti, T. Richmond, A. Baratoff, G. Borer, M. Dreier, M. Bernasconi, H.-J. Güntheroot: Europhys. Lett. **25**, 297 (1994)
63. R.P. Andres, T. Bein, M. Dorogi, S. Feng, J.I. Henderson, C.P. Kubiak, W. Mahoney, R.G. Osifchin, R. Reifenberger: Science **272**, 1323 (1996)
64. M. Dorogi, J. Gomez, R. Osifchin, R.P. Andres, R. Reifenberger: Phys. Rev. B **52**, 9071 (1995)
65. K.H. Park, J.S. Ha, W.S. Yun, M. Shin, K.-W. Park: Appl. Phys. Lett. **71**, 1469 (1997)
66. C.T. Black, D.C. Ralph, M. Tinkham: Phys. Rev. Lett. **76**, 688 (1996)
67. C.-S. Jiang, T. Nakayama, M. Aono: Appl. Phys. Lett. **74**, 1716 (1999)
68. J.G.A. Dubois, J.W. Gerritsen, S.E. Shafranjuk, E.J.G. Boon, G. Schmid, H. van Kempen: Europhys. Lett. **33**, 279 (1996)
69. E. Bar-Sadeh, Y. Goldstein, C. Zhang, H. Deng, B. Abeles, O. Millo: Phys. Rev. B **50**, 8961 (1994)
70. A.E. Hanna, M. Tinkham: Phys. Rev. B **44**, 5919 (1991)
71. M. Amman, R. Willkins, E. Ben-Jacob, P.D. Maker, R.C. Jaklevic: Phys. Rev. B **43**, 1146 (1991)
72. F. Negri, G. Orlandi, F. Zerbetto: Chem. Phys. Lett. **144**, 31 (1988)
73. N. Koga, K. Morokuma: Chem. Phys. Lett. **196**, 191 (1992)

74. D. Porath, O. Millo: J. of Appl. Phys. **85**, 2241 (1997)
75. J.K. Gimzewski, C. Joachim, R.R. Schlittler, V. Langlais, H. Tang, I. Johannsen: Science **281**, 531 (1998)
76. J.K. Gimzewski, C. Joachim: Science **283**, 1683 (1999)
77. C. Joachim, J.K. Gimzewski: Chem. Phys. Lett. **265**, 353 (1997)
78. R. Landauer: Nature **335**, 779 (1988)
79. L. Esaki: Phys. Rev. **109**, 603 (1958)
80. I.-W. Lyo, P. Avouris: Science **245**, 1369 (1989)
81. P. Avouris, I.-W. Lyo, F. Bozso, E. Kaxiras: J. of Vac. Sci. and Tech. A **8**, 3405 (1990)
82. P. Bedrossian, D.M. Chen, K. Mortensen, J.A. Golovchenko: Nature **342**, 258 (1989)
83. T. Yakabe, Z.-C. Dong, H. Nejoh: Appl. Surf. Sci. **121**, 187 (1997)
84. T. Komura, T. Yao: Phys. Rev. B **56**, 3579 (1997)
85. Y. Xue, S. Datta, S. Hong, R. Reifenberger, J.I. Henderson, C.P. Kubiak: Phys. Rev. B **59**, R7852 (1999)
86. N.D. Lang: Phys. Rev. B **55**, 9364 (1997)
87. K.F. Kelly, D. Sarkar, G.D. Hale, S.J. Oldenburg, N.J. Halas: Science **273**, 1371 (1996)
88. H.P. Lang, V. Thommen-Geiser, V. Frommerx, A. Zahab, P. Bernier, H.-J. Güntheroot: Europhys. Lett. **18**, 29 (1992)
89. J. Resh, D. Sarkar, J. Kulik, J. Brueck, A. Ignatiev, N.J. Halas: Surf. Sci. **316**, L1061 (1994)
90. N.D. Lang: Phys. Rev. B **34**, 5947 (1986)
91. N. Camillone III, T.Y.B. Leung, P. Schwartz, P. Eisenberger, G. Scoles: Langmuir **12**, 2737 (1996)
92. M.D. Porter, T.B. Bright, D.L. Allara, C.E.D. Chidsey: J. of the American Chemical Society **109**, 3559 (1987)
93. N.D. Lang, A.R. Williams: Phys. Rev. B **18**, 616 (1978)

5 Interaction Between Nonrelativistic Electrons and Optical Evanescent Waves

J. Bae, R. Ishikawa, and K. Mizuno

Scanning near-field optical microscopy (SNOM) has attracted much attention as a new optical technique to circumvent diffraction limits of the wavelength in conventional optics [1,2]. Near-field microscopes [3,4] have demonstrated subwavelength resolution of several tens of nanometers for imaging various kinds of objects, including semiconductors, superconductors, and biological tissue. In SNOM, near-field images of objects are produced by scattering the optical near-field on objects with a small probe tip such as a tapered optical fiber and a sharpened metal whisker. The following question always arises from the SNOM operating principle: how do SNOM images reflect optical properties of an object? The probe tip must be small compared to the object, not only to achieve higher resolution, but also to avoid strong disturbance of the near-field distribution. From this point of view, free electrons in vacuum are most suitable as a near-field probe, because electrons have the smallest size and therefore the least influence on the field.

Nonrelativistic electron beams in vacuum have been widely utilized for amplifying and oscillating coherent electromagnetic waves and for operating electron microscopes with subatomic resolution. Operating frequencies in these beam devices lie mostly in the microwave or millimeter wave region [5]. Much effort has been made to increase the operating speed of devices using the (inverse) Smith–Purcell effect [6,7], but the frequencies still remain in the submillimeter wave region [8,9]. Consequently, the interaction behavior between electrons and light, including quantum effects that occur at optical frequencies, is still under investigation.

In this chapter, we discuss the energy modulation of nonrelativistic electrons with evanescent waves contained in optical near-fields. In the next section, we present microgap interaction circuits that can be used at optical frequencies. In Sect. 5.2, a metal microslit among the microgap interaction circuits is adopted for interaction with an optical near-field and is discussed theoretically. Section 5.3 describes experiments performed at infrared wavelengths. The results demonstrate that the infrared laser can modulate the electron energy for a low energy electron beam using the microslit circuit. The extension of infrared investigations to the visible light region is discussed in Sect. 5.4. Section 5.5 reviews results presented in this chapter.

5.1 Microgap Interaction Circuits

5.1.1 Energy and Momentum Conservation

In the interaction between electrons and light, energy and momentum conservation must be satisfied [?]. Consider the case where an electron absorbs a photon. The dispersion relations for the electron and the photon are

$$W_{\mathrm{e}} = \sqrt{m^2c^4 + c^2{p_{\mathrm{e}}}^2}\ , \tag{5.1}$$

$$W_{\mathrm{p}} = \hbar\omega = cp_{\mathrm{p}}\ , \tag{5.2}$$

where W and p indicate the energy and the momentum, respectively, m is the electron rest mass, c is the speed of light, $\hbar = h/2\pi$, h is Planck's constant, and ω is the angular frequency. When W_{e} increases by W_{p}, the electron momentum p_{e} increases by

$$\Delta p = \frac{\hbar\omega}{v} \tag{5.3}$$

where v is the electron velocity. Since $p_{\mathrm{p}} = \hbar\omega/c$ is always smaller than Δp, momentum conservation is not satisfied and consequently the interaction between an electron and a photon never occurs in free space. For the interaction, some circuits that supply the shortfall in the electron momentum are indispensable.

Figure 5.1 shows possible interaction circuits having a smaller gap d than the wavelength:

(a) a thin dielectric film,
(b) a metal film gap,
(c) a metal microslit.

The dielectric film is the interaction circuit used by Schwarz and Hora in 1969 [11]. The metal film gap is very similar to conventional circuits used in klystrons [5]. The metal microslit was proposed for the study of electron-light interactions in an optical near-field [12].

In Fig. 5.1, the effective interaction length in the three circuits is limited so that an uncertainty exists in the momentum. This additional momentum p_{c} from the circuits is given approximately by

$$p_{\mathrm{c}} \sim \frac{h}{d} \tag{5.4}$$

If $p_{\mathrm{c}} > \Delta p$, momentum conservation can be satisfied for the interaction. In the case of the metal film gap, the condition for the gap d is derived from (5.3) and (5.4) as

$$d < \beta\lambda \tag{5.5}$$

where $\beta = v/c$ and λ is the wavelength of light. For the experimental parameters, $\lambda = 488$ nm and $\beta = 0.5$ which corresponds to an electron energy of about 80 keV, the gap must be narrower than 244 nm.

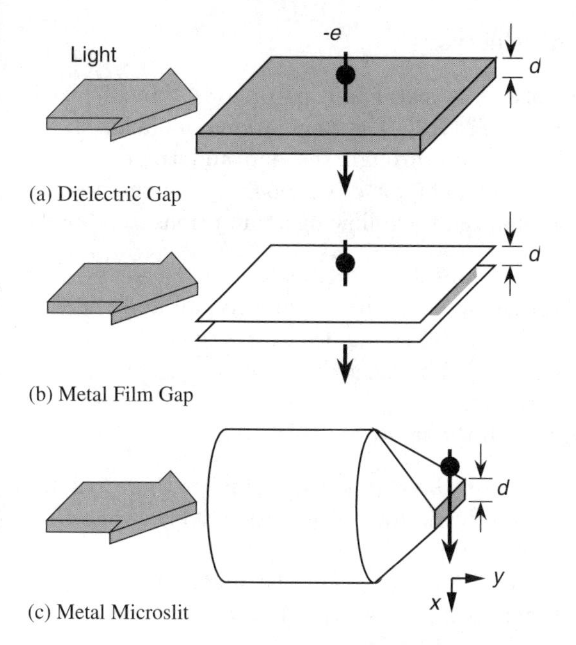

Fig. 5.1. Schematic diagram of three different gap circuits. The gap width d is smaller than the wavelength of the incident laser

The same considerations can be applied in a different way to the metal microslit shown in Fig. 5.1c, using an evanescent wave theory [12] rather than the uncertainty principle. A number of wave components with different wave numbers k_{ev} are induced in the slit by the incident light. On the basis of the evanescent wave theory, the wave numbers k_{ev} in the x-direction could extend from zero to infinity. From momentum conservation, k_{ev} must be equal to $\Delta p/\hbar$. Using (5.3), the following relationship between k_{ev} and v is found:

$$k_{\mathrm{ev}} = \frac{\omega}{v} \, . \tag{5.6}$$

Since k_{ev} is larger than the wave number $k_0 = \omega/c$ of light in free space, this wave is evanescent. This equation also represents phase matching between the electron and the optical evanescent wave, that is, $v_{\mathrm{p}} = v$, where v_{p} is the phase velocity of the evanescent wave given by $v_{\mathrm{p}} = \omega/k_{\mathrm{ev}}$. The validity of these theoretical considerations will be discussed in more detail later.

This type of metal microslit is more suitable than the other gap circuits for precise measurement of the electron–light interaction and for investigating quantum effects. This is because there are no disturbances caused by electron scattering in the metal film or the dielectric medium.

5.1.2 Transition Rates of Electrons

The transition rates of electrons in a metal film gap and a thin dielectric film have been estimated theoretically [12]. The calculational model is shown in Fig. 5.2. The electron beam passes through the gap and travels in the positive x-direction, whilst the laser light wave is propagated in the positive y-direction. To simplify the calculation, the following assumptions were made:

- the initial velocities of the electrons are same,
- propagation modes of the light wave in the metal film gap and the thin dielectric film are the TEM_{00} and TM_{01} modes, respectively,
- the incident light wave is polarized in the x-direction,
- the gap materials have no rf loss,
- the electrons interact only with the light.

The second assumption is valid when the gap width is smaller than $\lambda/2$ and $\lambda/(2\sqrt{n-1})$ for the metal and dielectric film gaps, respectively, where n is the refractive index of the dielectric film.

The transition rate for the electron energy in the metal film gap was calculated in accordance with analyses by Marcuse [14]. The calculated rate w_m is expressed as

$$w_\mathrm{m} = \frac{2qc\beta^2}{\varepsilon_0 \hbar^2 \omega^4} i P_\mathrm{i} \sin^2\left(\frac{\omega d}{2v}\right), \tag{5.7}$$

where q is the electron charge, i the electron current density, ε_0 the dielectric constant of free space, and P_i the power density of the incident light. The value of w_m represents the rate at which an electron absorbs one photon of energy $\hbar\omega$. The rate for emission of the photon is almost the same.

Similarly, it is possible to calculate a transition rate w_d for the dielectric film by quantizing the laser field of the fundamental TM_{01} propagation mode. The expression giving w_d is

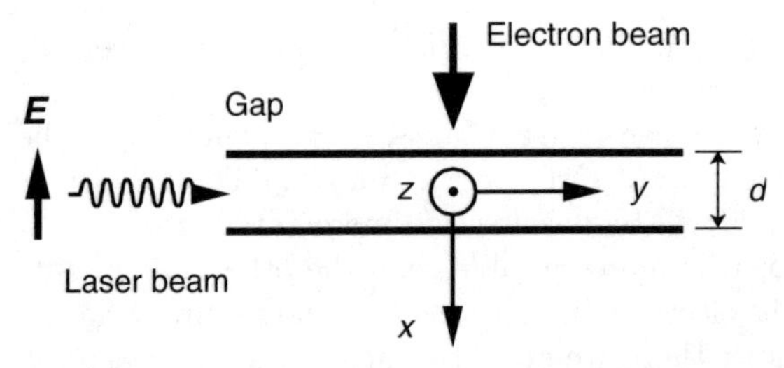

Fig. 5.2. Gap circuit configuration used for theoretical analysis

$$w_{\mathrm{d}} = \frac{qc}{\epsilon_0 \hbar^2 \omega^4} iP_{\mathrm{i}} \left| \frac{k_0 k_y d^2}{2{n_{\mathrm{i}}}^2 \cos(k_{\mathrm{i}x} d/2)} \right|^2$$

$$\times \frac{\left| 4 \frac{k_{\mathrm{i}x} d \sin(k_{\mathrm{i}x} d/2) \cos(\omega d/2v) - (\omega d/2v) \cos(k_{\mathrm{i}x} d/2) \sin(\omega d/2v)}{(k_{\mathrm{i}x} d)^2 - (\omega d/v)^2} \right|^2}{\left[\frac{k_0 d}{2n_{\mathrm{i}} \cos(k_{\mathrm{i}x} d/2)} \right]^2 + \left[\frac{({k_y}^2 - {k_{\mathrm{i}x}}^2)}{(n_{\mathrm{i}} k_{\mathrm{i}x})^2} - \frac{{n_{\mathrm{i}}}^2 ({k_y}^2 - {k_{\mathrm{e}x}}^2)}{(n_{\mathrm{e}} k_{\mathrm{e}x})^2} \right]}$$

$$\times \frac{1}{\frac{k_{\mathrm{i}x} d}{2{n_{\mathrm{i}}}^2} \tan(k_{\mathrm{i}x} d/2)} \tag{5.8}$$

where

$$k_{\mathrm{i}x}^2 + k_y^2 = (n_{\mathrm{i}} k_0)^2 \,, \quad k_{\mathrm{e}x}^2 + k_y^2 = (n_{\mathrm{e}} k_0)^2 \,,$$

$$|k_{\mathrm{e}x}| = \left(\frac{n_{\mathrm{e}}}{n_{\mathrm{i}}} \right)^2 k_{\mathrm{i}x} \tan k_{\mathrm{i}x} d/2 \,,$$

and n_{i} and n_{e} are the refractive indexes inside and outside the dielectric film, respectively, k_y is the wave number in the y-direction in free space, and $k_{\mathrm{i}x}$ and $k_{\mathrm{e}x}$ are the wave numbers in the x-direction inside and outside the film.

From (5.7) and (5.8), the transition rates can be estimated as a function of the gap width. Figure 5.3 shows the calculated results used to find the optimum gap width. The parameters used in the calculation are $\beta = 0.5$, $\lambda = 780\,\mathrm{nm}$, $n_{\mathrm{i}} = 1.45$ (SiO_2), and $n_{\mathrm{e}} = 1$. As can be seen from (5.6), w_{m} changes sinusoidally and peaks at the gap width satisfying the equation

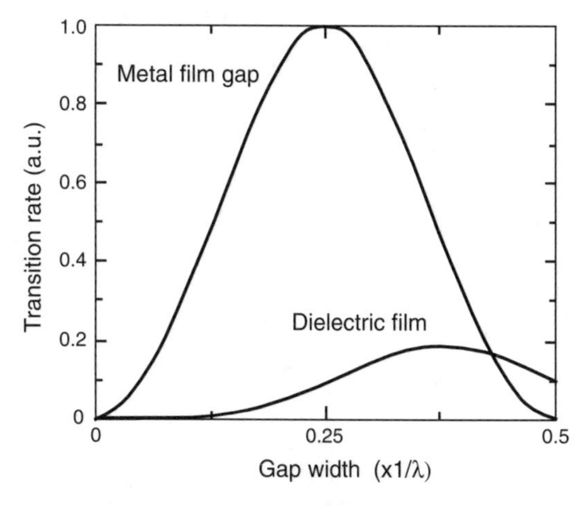

Fig. 5.3. Calculated transition rates for an electron with velocity $\beta = 0.5$ as a function of the gap width of the metal film gap and the SiO_2 dielectric film. The gap width is normalized to the laser wavelength

$$d = \beta\lambda\left(m + \frac{1}{2}\right), \tag{5.9}$$

where m is an integer. Substituting $\beta = 0.5$ and $m = 0$ for the first peak, an optimum gap width of $\lambda/4$ is obtained. The variation of w_{m} in Fig. 5.3 differs from the predictions of the klystron theory in which the maximum value of w_{m} occurs at $d \sim 0$ [5]. The difference comes from the different treatments of the photon density in the gap. The klystron theory assumes that the total number of photons stored in the gap is constant, whereas in our treatment it is assumed that the photon density in the gap is constant and determined by the incident laser power.

From Fig. 5.3, it can be seen that for the SiO_2 film, the first peak value of w_{d} is 0.18 times that of w_{m}. The optimum film thickness of 0.38 is also longer than the optimum width of the metal film gap. In the dielectric film, the laser field is distributed outside the film as an evanescent wave so that the number of photons inside the gap is smaller than that for the metal gap. The longer gap width increases the number of photons inside the dielectric film.

Figure 5.4 shows the calculated transition rates as a function of the light intensity for the metal film gap and the SiO_2 film. These transition rates represent the probability per unit time of one electron absorbing a photon. In the calculation, the optimum gap widths of 0.25λ and 0.38λ were used for the metal film gap and the SiO_2 film, respectively. Other calculation parameters take the same values as described previously. From Fig. 5.4, it can be seen that the transition rates are 1.1×10^{-2} and 2×10^{-3}/s for a power intensity of 10^6 W/cm^2 in the metal film gap and the SiO_2 film, respectively. The power

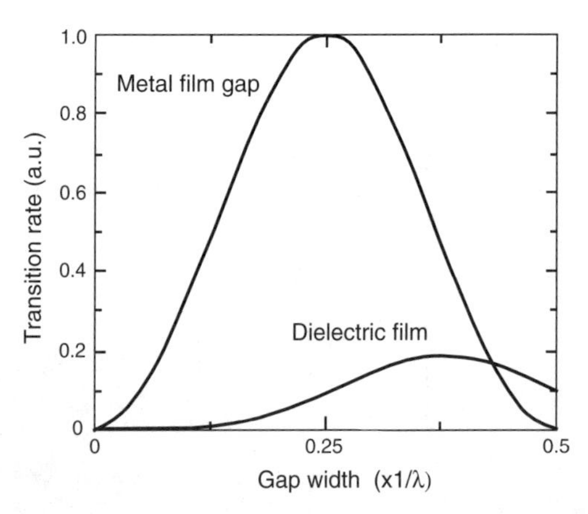

Fig. 5.4. Calculated transition rates as a function of the power intensity of the incident light for the metal film gap and the SiO_2 dielectric film at a wavelength of 780 nm

intensity of the laser corresponds to an output power of 790 mW focused onto an area of diameter 10 μm.

The transition rate just in front of the metal slit is intended to be the same as that of the metal film gap because the field distribution at the surface of the metal slit is similar to that in the film gap. A detailed analysis for the metal slit is described in a classical manner in the next section.

5.2 Metal Microslit

The metal microslit is used for generating optical near-fields in which an electron beam is modulated with light at optical frequencies. Optical near-field distributions on the slit were determined using the method of moments. The theory was checked by comparing with the measured field distributions at a microwave frequency. The energy changes of electrons in the near-field were evaluated numerically through computer simulations using the Lorentz force equations. From the results, the relationship between wavelength, slit width, and electron velocity was found.

5.2.1 Near-Field Distributions

The calculational model for a metal slit is shown in Fig. 5.5. In accordance with Chou and Adams' analysis [15], the electric field components E_x and E_y and a magnetic field component H_z in scattered waves from the slit were determined for an incident wave with field vectors $\boldsymbol{E}_{\mathrm{i}}$ and $\boldsymbol{H}_{\mathrm{i}}$. In the calculation, the following assumptions were made:

- a metal slit consists of two semi-infinite plane screens with perfect conductance and zero thickness,
- the incident wave is a plane wave with x-axis polarization and normal incidence upon the slit.

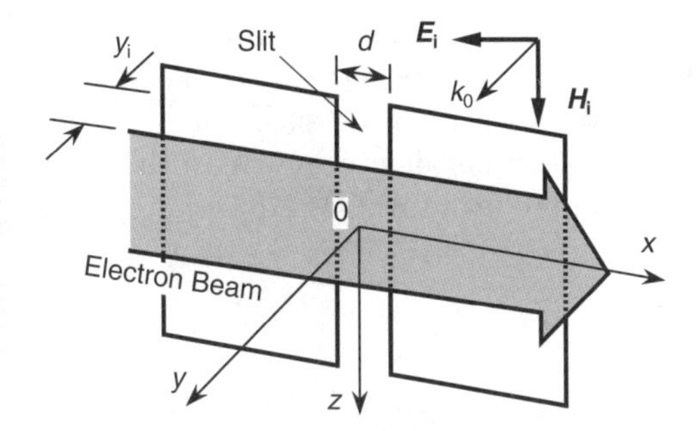

Fig. 5.5. Calculational model for the metal slit circuit

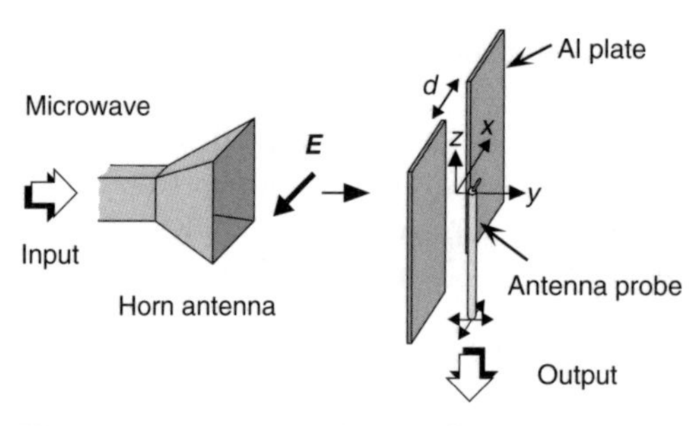

Fig. 5.6. Experimental setup for measuring the near-field distribution on the metal slit at 9 GHz

In order to check the theory, a near-field distribution on a metal slit was measured using a macrowave-scale model at 9 GHz (wavelength $\lambda = 33$ mm). The experimental setup is shown in Fig. 5.6. The metal slit consists of two aluminum plates with a height of 400 mm, a width of 190 mm, and a thickness of 1 mm. The antenna probe with a length of 1.6 mm was placed at the end of a thin coaxial cable with a diameter of 0.8 mm connected to a spectrum analyzer. This antenna detects an electric field intensity E_x in the x-direction, the dominant field in the interaction with electrons.

Figures 5.7a and b compare the calculated and measured field intensity distributions of E_x on the slit. The slit width d was 0.5λ. The field intensities were normalized to $E_{\mathrm{i}x}$ which is the measured field at $x = y = 0$ without the slit. In Fig. 5.7, the theoretical and experimental field distributions are quite similar except for small ripples appearing at $x/\lambda \sim -0.5$. The ripples could be caused by interference of waves scattered from the coaxial probe and the slit.

The theoretical and measured variations of E_x at $x = 0$ are plotted in Fig. 5.8. The theory for $d = 0.75\lambda$ agrees well with the measurement. When d decreases to 0.12λ, the deviation from the measurement increases due to the probe having a finite size of about 0.05λ. However, the theory accurately predicts the measured variations of the field intensities. These results indicate that the theory is valid within the bounds of experimental error.

From Fig. 5.8, it can be seen that the smaller the slit width, the steeper the field decay. The near-fields are localized within a distance of $y \sim d$ from the surface. These near-field distributions at the slit are similar to those for small apertures used as optical near-field probes in SNOM [16].

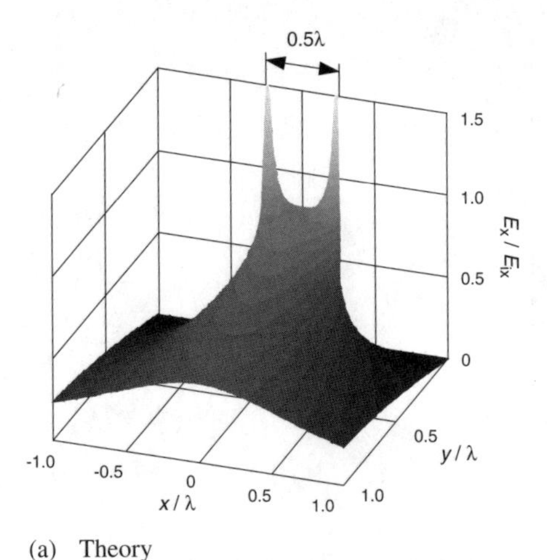

(a) Theory

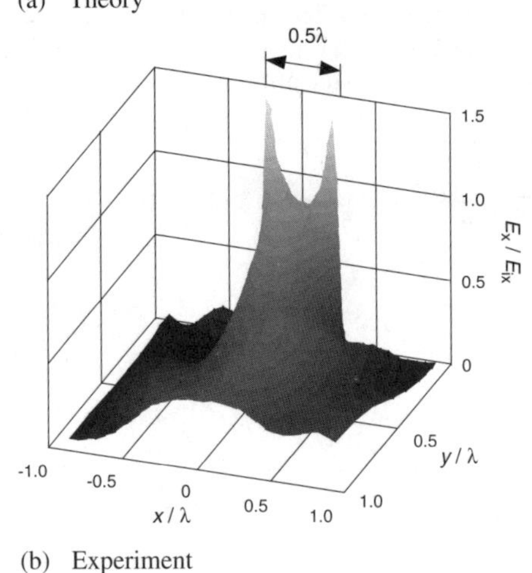

(b) Experiment

Fig. 5.7. (**a**) Calculated and (**b**) measured field distributions of the electric field component E_x in the x-direction on a slit of width $d = \lambda/4$. E_x is normalized to the incident field $E_{\mathrm{i}x}$ and the positions x and y are also normalized to the wavelength

5.2.2 Optimum Slit Width

Using the theoretical near-field distributions, energy changes of electrons passing close to the slit surface were estimated through computer simulation. The calculational model is the same as the previous one shown in Fig. 5.5. Electrons with velocity v_{i} move in the x-direction at distance y_{i} from the slit surface. All field components in the near-field, i.e., electric fields E_x, E_y and magnetic field H_z, were taken into account in the calculation. The total energy changes of the electrons were determined by integrating small energy changes with the Lorentz force over a small distance along the electron tra-

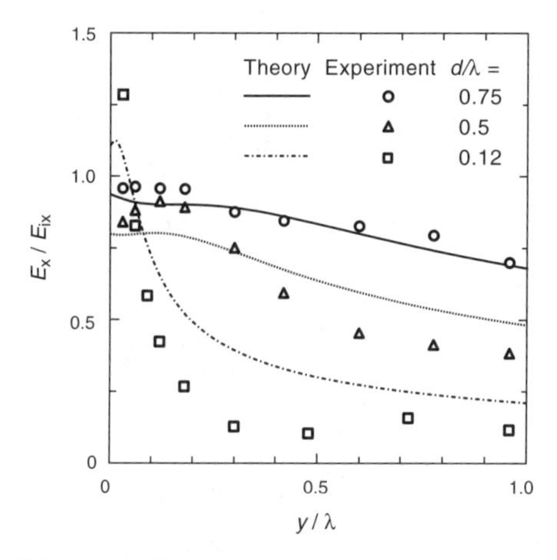

Fig. 5.8. Comparison between calculated and measured field intensities as a function of the distance y away from the slit surface for different slit widths at $x = 0$. The field intensities E_x are normalized to E_{ix} without the slit

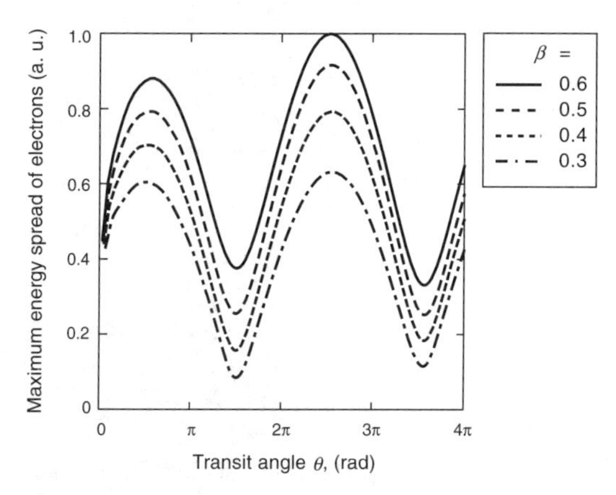

Fig. 5.9. Calculated maximum energy spread of electrons as a function of their transit angles θ for different electron velocities $\beta = v_i/c$ from 0.3 to 0.6 at $y_i = 0.01\lambda$. In the figure, the calculated energy spreads are normalized to the maximum value in the curve for $\beta = 0.6$ at $\theta \sim 2.5\pi$

jectory [17]. The integral length of ten times the slit width was chosen to cover the near-field region of the slit completely.

Maximum energy changes (spread) ΔW for electrons with different velocities β at $y_i = 0.01\lambda$ were calculated as a function of slit width d. Results are shown in Fig. 5.9. The abscissa is the transit angle $\theta = k_0 d/\beta$ corresponding to d. The ordinate is ΔW normalized to the peak in the curve for $\beta = 0.6$. In Fig. 5.9, the ΔW curves have two peaks, at $\theta \sim 0.5\pi$ and 2.5π. Note that these optimum transit angles do not depend on β. From the transition angles, it was found that optimum slit widths d exist for β and are given by

$$d = \beta\lambda \left(m + \frac{1}{4} \right) , \tag{5.10}$$

where m is an integer. Comparing (5.10) with (5.9), it follows that the optimum width in the slit is narrower than that in the metal film gap by $\beta\lambda/4$. This could be due to the non-uniform distribution of the near-field over the slit. In Fig. 5.9, the second peaks at wider widths are always larger than the first peaks at narrower widths. When the slit width becomes wider, the field intensity on the slit decreases as shown in Fig. 5.8 and the effective interaction length increases. The results thus indicate that the increase in the interaction length is a dominant effect for ΔW, at least up to $\theta = 2.5\pi$.

5.2.3 Phase Matching Condition

In Sect. 5.1.1, we said that electrons only interact with an evanescent wave contained in an optical near-field on a microslit. From energy and momentum conservation, the interaction condition, i.e., the phase matching condition between electron and evanescent wave, was derived as shown in (5.6). Let us check the validity of this condition using a computer simulation.

Figure 5.10 shows the calculated wave number (k) spectra of the near-fields E_x on the slits with $d/\lambda = 0.38$, 0.5, an 0.62 at $y_\mathrm{i} = 0.01\lambda$. The abscissa is the wave number k_x normalized to k_0 and the corresponding electron velocity calculated from (5.6). Amplitudes of wave components are normalized to the maximum value in the curve for $d/\lambda = 0.38$ at $k_x = 0$. The wave components with $k_x > k_0$ in Fig. 5.10 are evanescent waves that cannot propagate in free space because these waves have imaginary k_z.

In Fig. 5.10, the thick solid curve represents the calculated ΔW for the slit with $d/\lambda = 0.62$. In the calculation, a CO_2 laser with power density $10^8\,\mathrm{W/cm^2}$ at $\lambda = 10.6\,\mu\mathrm{m}$ was taken as the incident wave. This power density corresponds to a 10 kW output power focused onto an area of diameter 100 µm. As can be seen from Fig. 5.10, ΔW is proportional to the amplitude of the evanescent wave. Calculation results indicate that electrons with initial velocity v_i interact only with the evanescent wave of phase velocity $v_\mathrm{p}(= \omega/k_x) = v_\mathrm{i}$, so that the interaction condition derived from conservation theory is valid.

5.2.4 Interaction Space

Figure 5.11 shows the calculated ΔW for different electron velocities as a function of the position y from the slit surface. Slits having the optimum slit widths of $d/\lambda = 0.38$, 0.5, 0.62 for $\beta = 0.3$, 0.4, 0.5, respectively, were used for calculation. Other parameters used are the same as before.

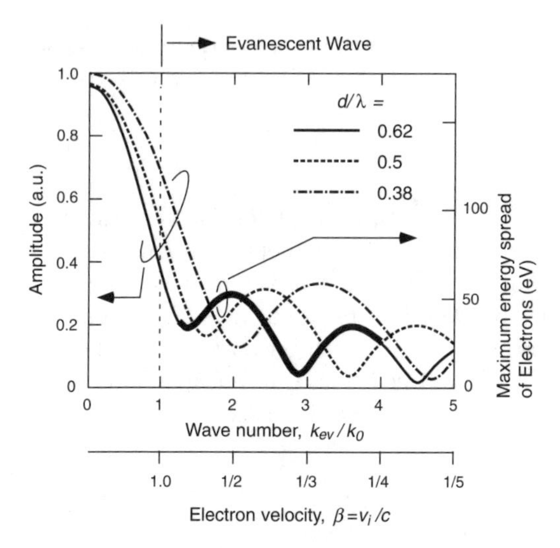

Fig. 5.10. Calculated wave number spectra for the near-field distributions of E_x on the slits with different widths $d/\lambda = 0.38$, 0.5, and 0.62. In the figure, k_x is the wave number in the x-direction and β is the electron velocity calculated by substituting k_x for k_{ev} in (5.5). The *thick curve* represents the maximum electron energy spread for different initial velocities from $\beta = 0.25$ to 0.8 which corresponds to an initial electron energy of 17–340 keV

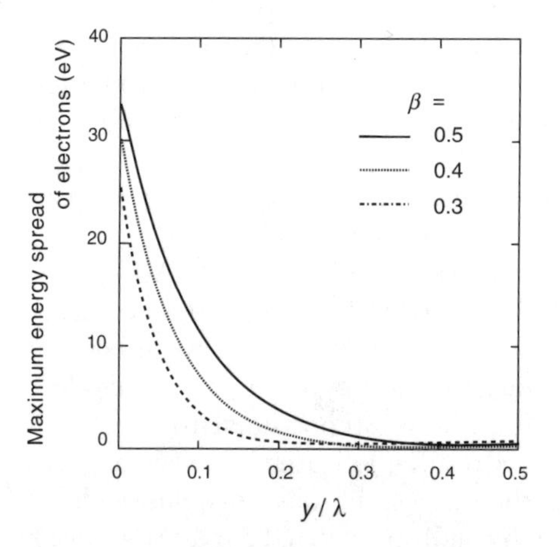

Fig. 5.11. Calculated maximum electron energy spreads as a function of y/λ for different electron velocities $\beta = 0.5$, 0.4, and 0.3

In Fig. 5.11, when y increases from zero to 0.5λ, ΔW falls off exponentially to near zero. Since ΔW is proportional to the field intensity at the slit, these curves represent effective field distributions over the slit. The shapes of the field distributions in Fig. 5.11 are quite different from those of the near-fields shown in Fig. 5.8. The field distributions are those of evanescent waves interacting with the electrons. The field decay constant α of the evanescent wave in the y-direction is given by

$$\alpha = k_0 \sqrt{\left(\frac{k_x}{k_0}\right)^2 - 1}\,. \tag{5.11}$$

This equation is easily derived from the relationship, ${k_0}^2 = {k_x}^2 + {k_y}^2$. Using (5.6) and (5.11), values of α for $\beta = 0.3, 0.4, 0.5$ are calculated to be $3.2k_0$, $2.3k_0$, $1.7k_0$, respectively. The decay constants agree with those estimated from the curves shown in Fig. 5.11.

The calculation results shown in Figs. 5.10 and 5.11 suggest that an electron beam could be used to measure a k-spectrum of an optical near-field appearing on a nano-object. This is a distinct feature of near-field microscopy with an electron beam as compared with conventional SNOMs.

The interaction space of the slit can be defined as $y_e = 1/\alpha$ because the field intensity of the evanescent wave falls off by e^{-1}. From (5.6) and (5.11), y_e is expressed as

$$y_e = \frac{\lambda}{2\pi}\frac{\beta}{\sqrt{1-\beta^2}}\,. \tag{5.12}$$

As can be seen from this equation, the interaction space in the slit circuit is strongly limited, particularly for a low energy electron beam. For $\beta = 0.5$ (~ 80 keV) and $\lambda = 10.6\,\mu$m, y_e is about 1 μm. The electrons must therefore pass very close to the slit surface to obtain significant energy exchanges with the light.

In practical measurements, the interaction space at the slit also depends on the field intensity of the evanescent wave. The energy changes of electrons are usually measured by an electron-energy analyzer with finite resolution. If the resolution is 1 eV, only electrons that gain or lose an energy of more than 1 eV can be detected. Therefore, the effective interaction space where detectable electrons are produced widens as the field intensity increases. For instance, the effective space in the curve for $\beta = 0.5$ in Fig. 5.11 is about 3.2 μm.

5.3 Modulation with Infrared Evanescent Waves

5.3.1 Experimental Setup

The metal microslit circuit was demonstrated experimentally at the infrared wavelength of 10.6 μm [18]. The experimental setup is shown in Fig. 5.12. An electromechanical Q-switched (EMQ) CO_2 laser [19] oscillates in TEM_{00} mode at $\lambda = 10.6\,\mu$m. This laser provides regular pulses with an output peak power of 11 kW and a width of 140 ns at a high repetition rate of 1 kpps. The laser beam focused through a ZnSe lens has a diameter of about 200 μm at

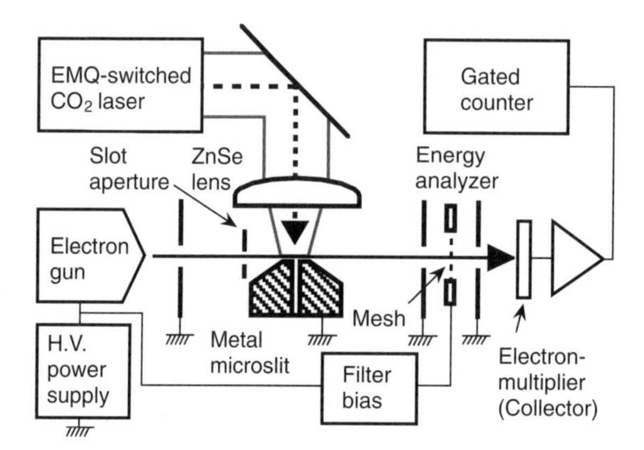

Fig. 5.12. Experimental setup for measuring the electron–light interaction in the infrared wave region

the slit surface. The metal slit consists of two polished copper blocks, and the width is 8.4 μm. The electron beam has a diameter of 240 μm. The slot aperture placed in front of the slit confines the beam area on the slit to 10 μm in height and 100 μm in width. The initial energy W_i of the electron beam was adjusted between 40 and 90 keV. The electron energy was measured using a retarding field analyzer [20].

The pulsed laser output modulates the energy of the electron beam so that the number of electrons through the analyzer changes during the pulse. The electrons passed through the energy analyzer were detected by a secondary electron multiplier (collector) connected to a gated counter which is triggered by the laser pulse.

5.3.2 Electron Energy Spectrum

Figures 5.13a and b show typical temporal changes in the laser pulse and the corresponding response from the gated counter for an electron beam with $W_i = 80$ keV. The collector current was $i = 1$ nA and the peak power of the laser was $P_i = 9$ kW. The output response was measured by the counter in box-car averager mode with a time resolution of 20 ns and integration time 2 s (accumulation number = 2 000). In this experiment, V_f was set to -5.5 V so that the counts in Fig. 5.13 represent the number of electrons gaining more energy than $|qV_f|$ from the laser.

As seen from Fig. 5.13, the shape of the output response differs considerably from that of the laser pulse. The response time of 380 ns in Fig. 5.13b is much longer than the laser pulse width of 140 ns. This output response results from the fact that the number of electrons interacting with the laser light is proportional to the laser field, not to the power. Note that a high signal-to-noise ratio (S/N) of more than 1 000 is achieved for the measurements.

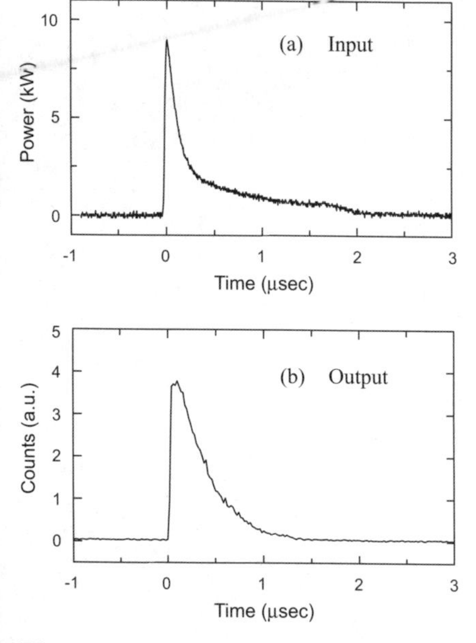

Fig. 5.13. Measured temporal changes of (**a**) the input laser pulse and (**b**) the output response from the gated counter for electrons with $W_\mathrm{i} = 80\,\mathrm{keV}$ at $\lambda = 10.6\,\mu\mathrm{m}$

Figure 5.14a shows the measured energy spectra of electrons with (A) and without (B) laser illumination, while Fig. 5.14b shows the difference between the two spectra $A - B$. In the experiment, $W_\mathrm{i} = 80\,\mathrm{keV}$ and $P_\mathrm{i} = 10\,\mathrm{kW}$ were used. The electron current was held below 1 pA to maintain a linear response from the electron multiplier even for a larger current input at $V_\mathrm{f} > 0$. To increase S/N in the measurement, a longer gate width of 1.5 μs was chosen and an integration time of 10 s was set in the counter.

From the measured spectrum B in Fig. 5.14a, it can be seen that the energy analyzer has resolution better than 0.8 eV for an 80 keV electron beam. The output count decreases gradually as V_f increases from +1 eV, due to the dispersion of the energy analyzer [20]. When the laser beam irradiates the electrons, the spectrum B changes to the spectrum A with a wider energy spread. Spectrum A still contains a number of electrons that have not interacted with the light. In order to remove these electrons and the dispersion effect from the measured spectrum A, the output counts in B were subtracted from those in A. Figure 5.14b thus indicates the energy spectrum only for the electrons that interacted with the light. From Fig. 5.14b, it can be seen that the 10 kW laser beam can give an energy spread of more than ±10 eV to the electrons.

Since the energy analyzer passes all the higher energy electrons, it is expected that for large V_f the counts of electrons with laser illumination

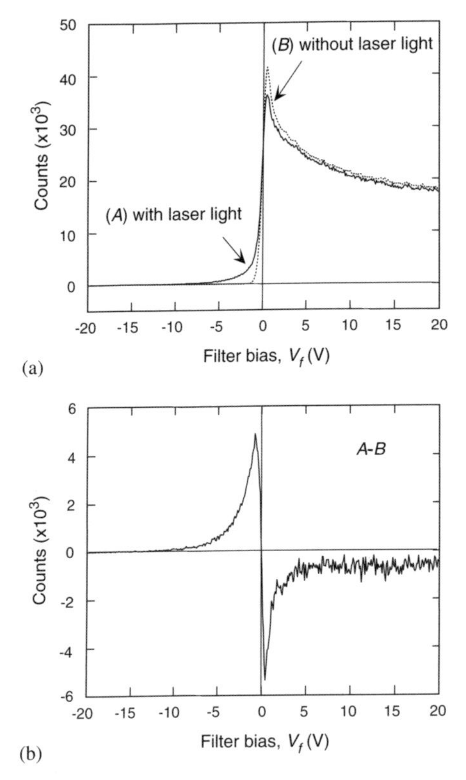

Fig. 5.14. Measured energy spectra for electrons with and without laser illumination (**a**) and the difference between the two spectra (**b**) for a laser beam with peak power 10 kW. The zero filter bias corresponds to the initial electron energy of 80 keV

should be the same as those for electrons without laser illumination. However, as shown in Fig. 5.14b, the counts with laser illumination are slightly smaller than those without laser illumination even at $V_f > +10$ V. This is due to deflection of the electron beam by the laser illumination. Consequently, a part of the electron beam has been clipped by the aperture before reaching the collector.

In Figs. 5.14a and b, about 42 000 electrons passed the slit, and about 9 000 electrons among them interacted with the laser beam. Since the height of the electron beam at the slit is 10 μm, this ratio of signal electrons to total electrons implies that the interaction space of the slit is about 2 μm, which agrees with the theoretical prediction described in Sect. 5.2.4.

5.3.3 Laser Field Dependence

In order to confirm that the electron energy spread is caused by the laser field on the slit, we measured the dependence of electron energy spreads on the

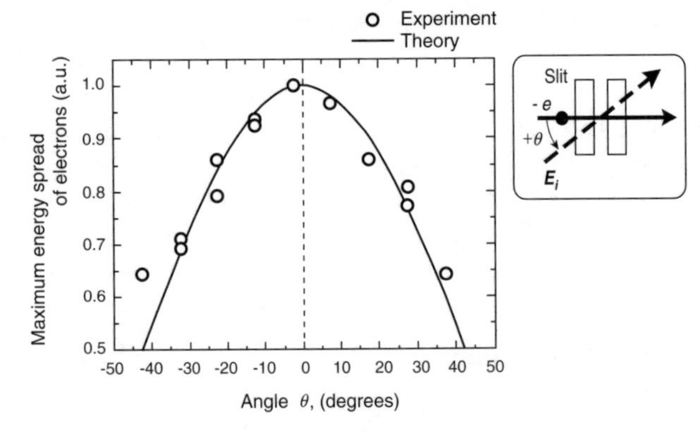

Fig. 5.15. Measured electron energy spreads versus polarization angle θ of the incident laser beam. The *solid curve* is the theoretically predicted change with θ. In the *inset*, $\boldsymbol{E}_\mathrm{i}$ is the electric vector of the laser beam

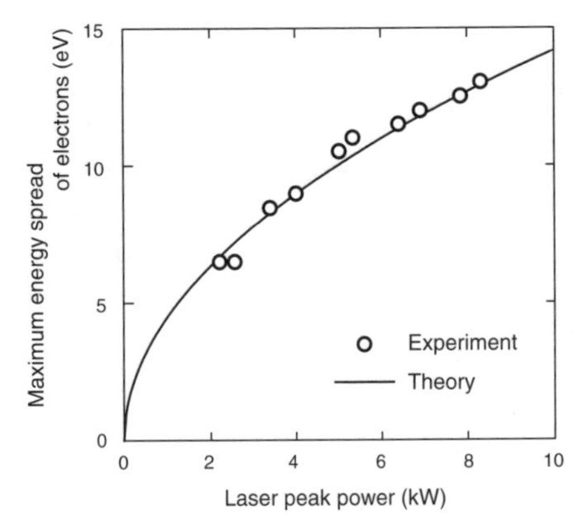

Fig. 5.16. Measured energy spreads of the electrons with an energy of 80 keV as a function of the peak power of the incident laser beam. The *solid curve* is the theoretical fit to the measured data

polarization of the laser beam and the power. The results for the polarization angles θ are shown in Fig. 5.15. The ordinate is the electron energy spread ΔW normalized to the maximum value at $\theta = 0$. The experimental parameters used are the same as the previous ones. The solid curve is the theoretical change in ΔW calculated assuming the relation $\Delta W \propto E_\mathrm{i} \cos\theta$, where E_i is the field intensity of the incident laser beam. Measurements agree well with theory.

Figure 5.16 shows the measured dependence of ΔW on the laser power. The solid curve is the theoretical fit to the measurements calculated by assuming that ΔW is proportional to the field intensity of the incident wave, i.e., the square root of the laser power.

In Fig. 5.16, the measured ΔW is 13 eV at $P_{\mathrm{i}} = 10$ kW, to be compared with the theoretical value of 22 eV estimated through the computer simulation. The reduction in the electron energy spread arises from differences between the actual slit and the ideal one used in the theoretical analysis. Since the actual slit consisted of two thick copper blocks with finite conductance, the amplitude of the evanescent wave may be small compared to the theoretical one. The results shown in Figs. 5.15 and 5.16 show that electrons are accelerated or decelerated by the infrared laser field at the microslit.

5.3.4 Electron Velocity Dependence

Figure 5.17 shows the measured and calculated energy spreads ΔW of the electrons as a function of the initial electron energy W_{i}. The initial electron velocities β corresponding to W_{i} are also indicated in the figure. The theoretical energy changes were calculated for a slit width of 7.2 μm to obtain the best fit to the measurements. In the figure, ΔW was normalized to the maximum values of 15 eV for the measurement and 34 eV for the theory, with $W_{\mathrm{i}} = 90$ keV. Due to the limitations of the electron gun, the measurements were made with W_{i} above 40 keV.

As described in Sect. 5.2.3, the measured variation of the electron energy spread shown in Fig. 5.17 represents the wave number spectrum of the optical near-field at the slit. The measured wave number spectrum implies that the effective width of the slit is not 8.4 μm as measured by the optical microscope, but 7.2 μm for the laser wave at $\lambda = 10.6$ μm.

As can be seen from Fig. 5.17, electrons with lower energy, less than 50 keV, were modulated by the laser beam in the microslit. The electron

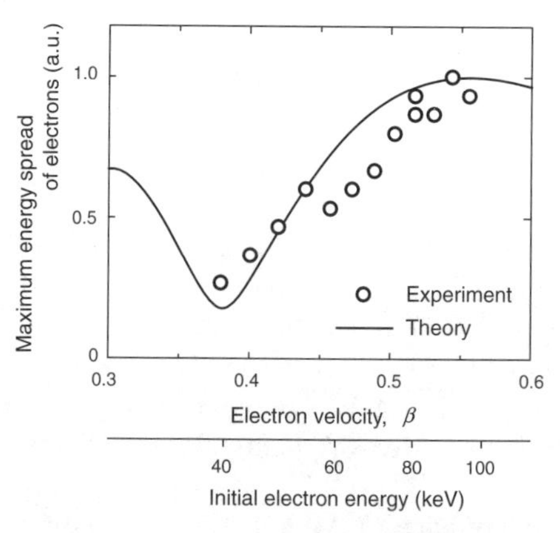

Fig. 5.17. Measured electron energy spreads as a function of the initial electron energy. In the figure, the electron velocities β corresponding to the electron energy are also indicated. The *solid curve* is the theoretically predicted electron energy spread for the slit with width 7.2 μm

energy spread is about 4 eV at $W_\mathrm{i} = 40\,\mathrm{keV}$ which is still enough to be able to measure the interaction. These results indicate that nonrelativistic electrons can be modulated using the metal microslit at optical frequencies, as predicted by theory.

5.4 Interaction with Visible Light

Figure 5.18 shows a conceptual drawing of the experimental system for the interaction of nonrelativistic electrons with light at shorter wavelengths than about 1.2 μm. The metal microslit is fabricated at the end of an optical fiber so that the laser beam is guided to the slit without any precise adjustment. Since light with shorter wavelengths has a photon energy higher than 1 eV, and the electron energy analyzer can resolve the energy changes of the electrons due to the photons, a quantum step in the electron energy is detectable.

On the basis of the theoretical and experimental results, the number of signal electrons in the interaction was estimated in both classical and quantum treatments, assuming a laser with $\lambda = 780\,\mathrm{nm}$ and output power of 30 mW, and an electron beam with velocity $\beta = 0.5$ and density $1\,\mathrm{mA/cm^2}$. It was also assumed that the diameter of the laser beam in the fiber was 6 μm.

According to the classical theory described in Sect. 5.2, the energy change of the electrons is estimated to be less than 0.1 eV. This energy change is too small to allow observation of the interaction in practical experiments. In order to give more than 1 eV to the electrons, a higher output power than 1 W is required for the laser. However, this laser power is available in experiments.

In contrast to the conclusion from classical considerations, a different result is derived from a quantum mechanical treatment. Assuming the transition rates in the metal film gap and the microslit to be the same, the transition rate of about 1×10^{-3}/s, which is the rate for an electron to absorb a

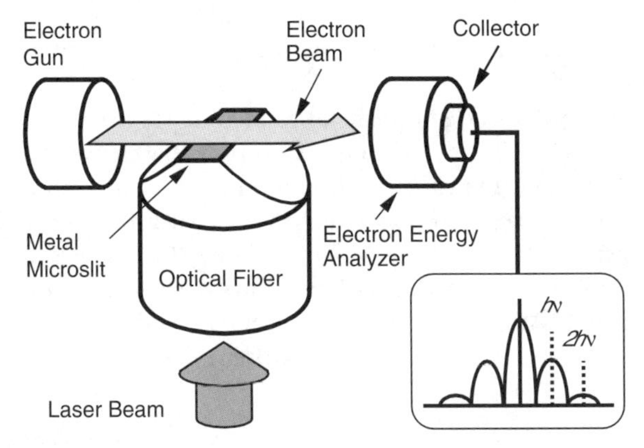

Fig. 5.18. Conceptual drawing of the experimental setup for measuring the electron–light interaction in the near-infrared and visible light regions

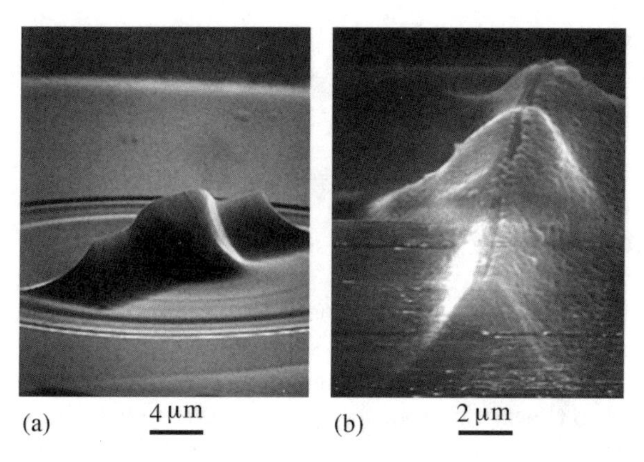

Fig. 5.19. Metal microslit fabricated at the center of the end of a single-mode optical fiber. (**a**) Ridge structure with a taper angle of 80°, (**b**) metal slit with a width of 270 nm fabricated on the ridge structure

laser photon, was obtained using (5.7). From (5.10) and (5.12), the optimum slit width and the interaction space are 490 and 72 nm, respectively. Then 2.7×10^7 electrons per second can pass through the space of 72 nm × 6 μm at the slit. The number of signal electrons is thus more than 20 000 particles per second. This number is enough to observe the electron–light interaction experimentally. Since the interaction behavior between electron and light in a transition region from the classical to quantum regimes is still vague, the final conclusion for the above discussions should be drawn from experimental evidence.

In the experiment at visible light wavelengths, a key device is the metal microslit with submicron width. The microslit is fabricated at the top of the ridge, as shown in Fig. 5.18. This ridge structure is needed to avoid a collision of the electrons with the slit due to the image force acting on the electrons in the proximity of the slit surface. Such a microslit can be fabricated using a chemical etching technique in a similar way to conventional near-field probes with nanometric apertures [21,22].

Figure 5.19 shows a prototype of the microslit fabricated at the center of the core of an optical fiber with a 125 μm diameter. Figure 5.19a shows the ridge structure and Fig. 5.19b the same ridge with aluminum coating and a 270 nm gap at the top. The ridge structure has a length of 5.5 μm for a core diameter of 8.8 μm, a height of 7.2 μm, and a taper angle of 80°. Both the flatness along the length of the ridge and the radius of curvature at the top of the ridge are less than 30 nm. Another probe with a 180 nm wide slit was also fabricated using gold instead of aluminum as a coating metal.

5.5 Conclusion

Energy modulation of an electron beam with light waves was analyzed theoretically using classical and quantum treatments. Three gap circuits, a metal film gap, a dielectric film, and a metal microslit were discussed as possible interaction circuits. Theory showed that the transition rates of electrons in the gap circuits were high enough to demonstrate the electron–light interaction experimentally. To achieve the interaction with an optical near-field, the metal microslit structure was adopted as the interaction circuit. Classical theory revealed that electrons passing through the slit can interact with an evanescent wave contained in the near-field induced on the microslit by light illumination. The relationship between the slit width, electron velocity, and light wavelength for the strongest interaction was determined through computer simulation. The microslit interaction circuit was successfully demonstrated experimentally at the infrared wavelength of 10.6 μm. Experiments showed that the energy of electrons can be changed by more than 10 eV with 10 kW laser illumination for a nonrelativistic electron beam with low energy, less than 50 keV. These results provide experimental backing for theoretical predictions. From the theoretical and experimental results, it was concluded that an optical near-field on a small object can be detected using an electron beam, so that a new type of near-field microscopy with free electrons may be feasible.

References

1. D.W. Pohl, W. Denk, and M. Lanz: Optical Stethoscopy: Image Recording with Resolution $\lambda/20$. Appl. Phys. Lett. **44**, 651–653 (1984)
2. A. Harootunian, E. Betzig, M. Isaacson, and A. Lewis: Super-resolution Fluorescence Near-Field Scanning Optical Microscopy. Appl. Phys. Lett. **49**, 674–676 (1986)
3. E. Betzig and J.K. Trautman: Near-Field Optics: Microscopy, Spectroscopy, and Surface Modification Beyond the Diffraction Limit. Science **257**, 189–195 (1992)
4. R.U. Maheswari, H. Tatsumi, Y. Katayama, M. Ohtsu: Observation of Subcellular Nanostructure of Single Neurons with an Illumination Mode Photon Scanning Tunneling Microscope. Opt. Commun. **120**, 325–334 (1995)
5. R.G.E. Hutter: *Beam and Wave Electronics in Microwave Tubes*, D. Van Nostrand Co. Inc., Toronto (1960)
6. S.J. Smith and E.M. Purcell: Visible Light from Localized Surface Charges Moving across a Grating. Phys. Rev. Lett. **92**, 1069 (1953)
7. K. Mizuno, S. Ono, and O. Shimoe: Interaction Between Coherent Light Waves and Free Electrons with a Reflecting Grating. Nature **253**, 184–185 (1975)
8. J. Bae, H. Shirai, T. Nishida, T. Nozokido, K. Furuya, and K. Mizuno: Experimental Verification of the Theory on the Inverse Smith–Purcell Effect at a Submillimeter Wavelength. Appl. Phys. Lett. **61**, 870–872 (1992)
9. J. Urata, M. Goldstein, M.F. Kimmitt, A. Naumov, C. Platt, and J.E. Walsh: Superradiant Smith–Purcell Emission. Phys. Rev. Lett. **80**, 516–519 (1998)

10. R.H. Pantell: Interaction Between Electromagnetic Fields and Electrons. In: R.A. Carrigan and F.R. Huson (Ed.) AIP conference proceedings No. 87, *Physics of High Energy Particle Accelerators*. American Institute of Physics, 863–918 (1981)
11. H. Schwarz and H. Hora: Modulation of an Electron Wave by a Light Wave. Appl. Phys. Lett. **15**, 349–351 (1969)
12. J. Bae, S. Okuyama, T. Akizuki, and K. Mizuno: Electron Energy Modulation with Laser Light Using a Small Gap Circuit – A Theoretical Consideration. Nucl. Instrum. & Methods in Phys. Research A **331**, 509–512 (1993)
13. G.A. Massey: Microscopy and Pattern Generation with Scanned Evanescent Waves. Appl. Opt. **23**, 658–660 (1984)
14. D. Marcuse: *Engineering Quantum Electrodynamics*, Academic Press Inc., New York (1970) pp. 127–142
15. T.Y. Chou and A.T. Adams: The Coupling of Electromagnetic Waves Through Long Slots. IEEE Trans. Electromagnetic Compatibility **EMC-19**, 65–73 (1977)
16. Y. Leviatan: Study of Near-Zone Field of a Small Aperture. J. Appl. Phys. **60**, 1577–1583 (1986)
17. J. Bae, K. Furuya, H. Shirai, T. Nozokido, and K. Mizuno: The Inverse Smith–Purcell Effect in the Submillimeter Wave Region – Theoretical Analyses. Jap. J. Appl. Phys. **27**, 408–412 (1988)
18. J. Bae, R. Ishikawa, S. Okuyama, T. Miyajima, T. Akizuki, T. Okamoto, and K. Mizuno: Energy Modulation of Nonrelativistic Electrons with a CO_2 Laser Using a Metal Microslit. Appl. Phys. Lett. **76**, 2292–2294 (2000)
19. J. Bae, T. Nozokido, H. Shirai, H. Kondo, and K. Mizuno: High Peak Power and High Repetition Rate Characteristics in a Current Pulsed Q-Switched CO_2 Laser with a Mechanical Shutter. IEEE J. Quantum Electron. **30**, 887–892 (1994)
20. J.F. Graczyk and S.C. Moss: Scanning Electron Diffraction Attachment with Electron Energy Filtering. Rev. Sci. Instrum. **40**, 424–433 (1969)
21. T. Pangaribuan, S. Jiang, and M. Ohtsu: Two-Step Etching Method for Fabrication of Fiber Probe for Photon Scanning Tunnelling Microscope. Electron. Lett. **29**, 1978–1979 (1993)
22. M. Ohtsu: Progress of High-Resolution Photon Scanning Tunneling Microscope Due to a Nanometric Fiber Probe. J. Lightwave Technol. **13**, 1200–1221 (1995)

6 The Tunneling Time Problem Revisited

N. Yamada

6.1 What is the Issue?

Tunneling is undoubtedly the most important quantum phenomena with a wide range of applications. Like other physical phenomena, it should be possible to understand tunneling phenomena from a spatial and also from a temporal point of view (see Fig. 6.1). Of interest from a spatial point of view is the question as to how many particles initially on one side of the barrier will finally be found on the other side of the barrier at sufficiently later times. We are thus led to the concept of tunneling probability. Of interest from a temporal point of view would be the tunneling time, that is, the time a particle 'spends' in the barrier region. From a spacetime point of view, it seems natural to characterize tunneling phenomena according to the two quantities tunneling probability and tunneling time.

The lifetime is well known as a characteristic time scale associated with tunneling. The tunneling time must not be confused with the lifetime (see Fig. 6.2). When a system tunnels out of a metastable state, for example, the lifetime is intuitively the measure of the time during which the particle is trapped in the potential well (classically allowed region), while the tunneling time is thought to be the time that the particle is under the potential barrier (classically forbidden region).

In applying tunneling phenomena, only the tunneling probability has been utilized. It would be safe to say that the tunneling time plays no role today in tools and devices based on tunneling. This raises the expectation that something new could be built or the performance of existing tunneling devices

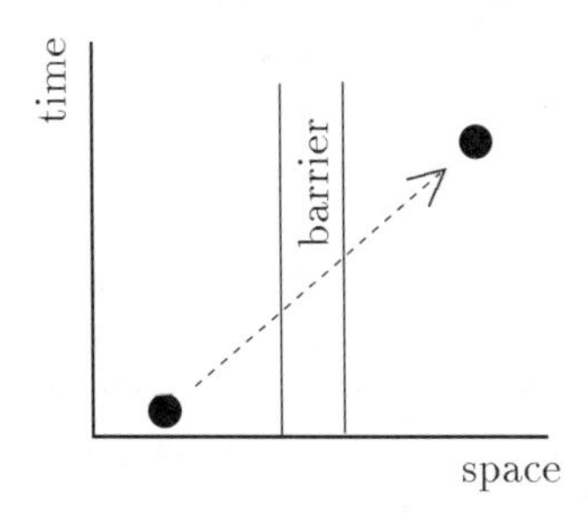

Fig. 6.1. Space–time picture of tunneling

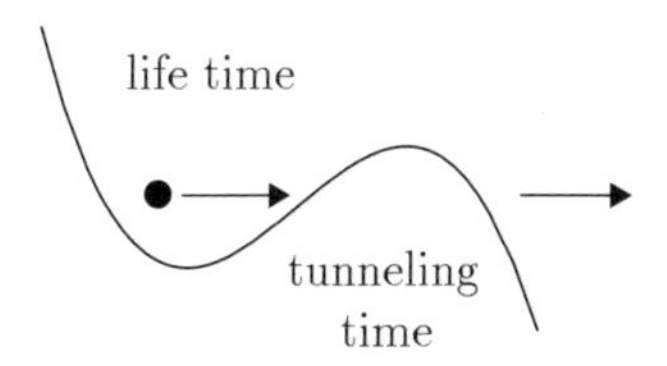

Fig. 6.2. Tunneling out of a metastable state: lifetime and tunneling time

could be improved if the tunneling time were utilized to control tunneling phenomena. This indicates the technological importance of tunneling time. Another point is that the tunneling spectroscopy as it exists today extracts information from the relevant system by measuring the tunneling probability (or tunneling current). If one could find a way to probe the system via a properly designed tunneling time measurement, a new area of research that deserved the name 'tunneling-time spectroscopy' would appear and the traditional one would then be called 'tunneling-probability spectroscopy'. This is where we find the importance of the study of tunneling time from an academic point of view. Scientists have been utilizing only one aspect of tunneling and that the utilization of another aspect, the time of tunneling, seems to be a fascinating direction for tunneling-related science and technology in the twenty-first century.

The barrier to overcome is high, however. There is no agreement as to what the tunneling time is and even as to the meaning of the concept of tunneling time. Regarding tunneling probability, its *concept* is very clear as described above, though the actual calculation of the probability is not always easy. On the other hand, the concept of tunneling time is not clear despite extended debate over the past 70 years. Therefore, the first thing to be done is to put an end to the long dispute and establish the concept of tunneling time. The purpose of this chapter is to present the problem of tunneling time for those readers who routinely deal with tunneling phenomena in their research but only in terms of tunneling probability. It is hoped that this chapter will inspire such readers to begin thinking about tunneling time.

In Sect. 6.2, we survey a variety of ideas that have been proposed to define and/or measure the tunneling time (see, for example, [1–8] for more detailed reviews of the problem). We will see that different methods of defining/measuring the time lead to different answers. All these methods are based on physically plausible ideas and often give reasonable answers when applied to non-tunneling regimes such as free propagation or transmission over a barrier or well. This implies that such 'physically plausible ideas' were reached by implicitly imaging the dynamics in non-tunneling regimes where classical mechanical intuition based on a particle picture works. However, tunneling is a purely quantum mechanical phenomena where the validity of such an intuition is no longer warranted. We must therefore treat the problem in a way securely rooted in the foundations of quantum mechanics.

A pessimist, being tired of repeated arguments with no convergence in the results, would argue that it does not make sense to talk about tunneling time. It should make sense, however, simply because tunneling is a physical phenomena that occurs, like other physical phenomena, in space–time. In my opinion, an approach that is free from unwarranted 'plausible arguments' but is instead securely rooted in the foundations of quantum mechanics would change this controversial situation. The author reported an effort in this direction in [9]. Section 6.3 gives an overview of this approach, which is intended for those who were once interested in the tunneling time problem but have stopped thinking about it, under the negative impression that the problem is not meaningful.

The question asked in Sect. 6.3 is more fundamental than the questions asked in most other approaches. That is, the definability of the probability distribution of tunneling time[1] is questioned and investigated, while many other arguments implicitly *assume* the existence of the distribution and try to construct it in a plausible way. Such an assumption does not seem to be founded. What must be clarified first is the definability of the distribution. Until recently, however, it was not clear how one could make such a clarification in quantum mechanics. It will be argued here that the consistency criterion provided by the consistent (or decoherent) histories approach to quantum mechanics is best suited to the purpose.

The consistent histories approach (CHA) is a general theory that deals with the question as to whether or not quantum probabilities are definable for a given set of alternatives.[2] CHA was developed by Griffiths [10], Omnés [11,12], Gell-Mann and Hartle [13,14], and also by the present author with Takagi [15,16].

As we shall see, a negative result follows from the consistency criterion, i.e., the probability distribution is not definable. The implicit assumption behind many arguments is therefore incorrect. We should be able to convince ourselves of this negative result by recalling the situation we face in the two-slit experiment. Since the alternative that the particle goes through the upper slit and the other alternative that it goes through the lower slit interfere,[3]

[1] Quantum mechanically, it is unlikely that the value of tunneling time is a constant that is uniquely determined from the initial condition of the particle and the barrier shape. Rather, a distribution of the values of tunneling time seems to be more natural to consider.

[2] Those who have studied CHA will notice that the equations that appear in Sect. 6.3 do not look like CHA equations. This is because, unlike most CHA studies where the relevant alternatives are defined by the products of projection operators at different times, Feynman paths are used to define alternatives in the present study.

[3] The two alternatives, which are classically mutually exclusive, are said to be *interfering alternatives*.

probabilities cannot be defined for the two alternatives in a consistent way.[4] Likewise, the alternative that the particle spends a certain amount of time in the barrier region and another alternative that it spends another amount of time in the region interfere,[5] making it impossible to define probabilities for the alternatives in a consistent way.

The role of CHA in the above story is to provide a rigorous definition for the measure of the interference (decoherence functional). What the author has found using the measure is that interference never vanishes in tunneling regimes. This negative conclusion immediately raises the following question: what can then be said about tunneling time? This is the second question treated in Sect. 6.3. The answer given there is that the *range* of values of tunneling time is 'speakable'. This is also understandable in analogy with the 'one-slit experiment'. Readers not interested in technical details can skip all equations in Sect. 6.3 but are instead encouraged to understand the results with the help of the two analogies employed there (see Fig. 6.7). The author does not claim to have solved the long-standing problem. Rather, the problem is simply redefined and a new starting point is set for constructive discussion. What physical results are obtained from this new starting point is mostly left to future studies. It is worth repeating, however, that the most important thing for the tunneling time problem is to have a sensible definition of the problem, and such a definition has, I believe, been presented. The long history of controversy does not mean that the tunneling time problem is meaningless.

Throughout the text, our analysis is restricted to the case of a rectangular potential barrier in one-dimensional space. Since our interest is a conceptual issue, this would be a reasonable choice. As stressed by Støvneng [17], of course, a fundamental study is not meaningful if it cannot be related to experiment, and ultimately to applications. Sadly, it is too early to speak of applications of tunneling time. For the moment we can only focus on conceptual issues. Once a foundation is established, applications should become possible. For most readers, this will be frustrating, but we should recall that the Esaki diode was invented in 1957 and STM in 1982, while tunneling phenomena were already known in the 1920s.

6.2 Well-known Approaches

6.2.1 Notation

There are many theoretical ideas for defining tunneling time. Only a few of these will be described here. Since the derivation of all the details is not the purpose here, intermediate equations will be omitted as far as possible.

[4] In such a way that the superposition principle for amplitudes (wave nature) and the sum rule for probabilities (particle nature) do not conflict.

[5] This interference may be said to be a temporal interference, while the interference in the two-slit experiment is a spatial interference.

To avoid complication, descriptions will be made for a square barrier in one dimension, although some of the ideas have been analyzed in more general circumstances in the literature. The following descriptions are not necessarily faithful to the original notations.

The square barrier of height V_0 is assumed to cover the region $a < x < b$. The Hamiltonian of the particle is simply

$$H = -\frac{\hbar^2}{2m}\frac{\partial^2}{\partial x^2} + V_0\,\Theta_{ab}(x)\,, \tag{6.1}$$

where m is the mass of the particle and

$$\Theta_{ab}(x) \equiv \begin{cases} 1 & \text{for } a < x < b\,, \\ 0 & \text{otherwise}\,. \end{cases} \tag{6.2}$$

Some approaches use wave packets and others use plane waves. In either case, what is basic to the analysis is the normalized eigenfunction $u_k(x)$ which takes the following form outside the barrier:

$$u_k(x) = \begin{cases} \dfrac{1}{\sqrt{2\pi}}\left(\mathrm{e}^{\mathrm{i}kx} + R\mathrm{e}^{-\mathrm{i}kx}\right) & \text{for } x < a\,, \\ \dfrac{1}{\sqrt{2\pi}} T\mathrm{e}^{\mathrm{i}kx} & \text{for } x > b\,, \end{cases} \tag{6.3}$$

where $k > 0$, $d \equiv b-a$, and the complex coefficients R and T are the reflection and transmission amplitudes, respectively, satisfying $|R|^2 + |T|^2 = 1$. Only T is important in our analysis. It is given by

$$T = \mathrm{e}^{-\mathrm{i}kd}\left[\cos\left(d\sqrt{k^2 - k_{V_0}^2}\right) - \mathrm{i}\,\frac{k^2 - k_{V_0}^2/2}{\sqrt{k_2 - k_{V_0}}}\sin\left(d\sqrt{k^2 - k_{V_0}^2}\right)\right]^{-1} \tag{6.4}$$

with $k_{V_0} \equiv \sqrt{2mV_0}/\hbar$. The decomposition of T into its modulus and phase is often useful:

$$T = |T|\,\mathrm{e}^{\mathrm{i}\theta} = |T|\,\mathrm{e}^{\mathrm{i}\phi}\mathrm{e}^{-\mathrm{i}kd}\,. \tag{6.5}$$

For a particle described by a wave packet $\Psi(x,t)$ (initially localized in $x < a$ and moving towards the barrier), the tunneling (or transmission) probability P_{t} is given by

$$P_{\mathrm{t}} \equiv \lim_{t\to\infty}\int_b^\infty \mathrm{d}x\,|\Psi(x,t)|^2 = \int \mathrm{d}k\,|\varphi(k)|^2\,|T(k)|^2 \tag{6.6}$$

$$\to |T(k_0)|^2 \text{ as } |\varphi(k)|^2 \to \delta(k - k_0)\,, \tag{6.7}$$

where $\varphi(k)$ is the k-space wave function, which is assumed to vanish for $k < 0$ (to make a right-moving wave packet) and also for $k > \sqrt{2mV_0}/\hbar$ (otherwise the particle can go over the barrier).[6] Equation (6.7) gives the tunneling probability for the stationary case with wave number k_0. The reflection probability P_{r} can be calculated from $P_{\mathrm{r}} = 1 - P_{\mathrm{t}}$.

[6] Note that the tunneling probability is originally defined as the probability of finding a particle somewhere on the right side of the barrier at sufficiently later

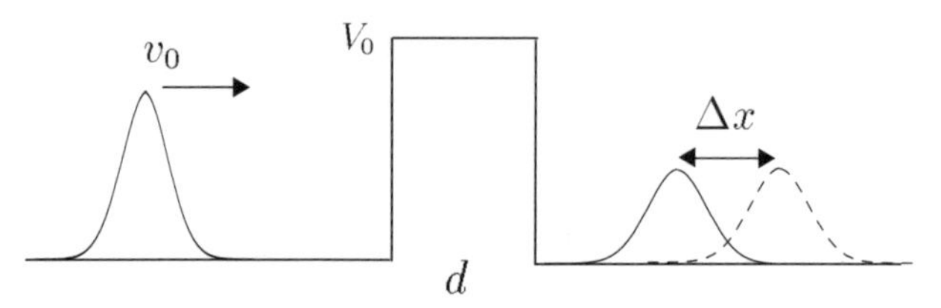

Fig. 6.3. Tunneling time from the motion of wave packets

6.2.2 Following the Motion of a Wave Packet

This is probably the simplest idea that comes to mind from a particle point of view. As an incident wave packet, we consider a wave packet that is narrow in k-space and thus wide in real space. This corresponds to a particle of nearly monochromatic kinetic energy. Explicitly,

$$\Psi(x,0) = \int \frac{\mathrm{d}k}{\sqrt{2\pi}}\, \psi(k-k_0)\, \mathrm{e}^{\mathrm{i}kx}\,, \tag{6.8}$$

where ψ is such that $|\psi(k-k_0)|$ is sharply peaked around $k = k_0 > 0$ and satisfies $\int \mathrm{d}k\, |\psi|^2 = 1$. At the initial time $t = 0$, the incident packet is supposed to be localized well to the left of the barrier and is moving towards the barrier with its peak velocity $v_0 = \hbar k_0/m$. At sufficiently later times, the wave function is made up of two spatially non-overlapping parts, the reflected and the transmitted packets. The reflected packet has a clear peak in $x \ll a$ that leaves the barrier with the velocity $-v_0$, while the transmitted packet has a clear peak in $x \gg b$ leaving the barrier with the velocity v_0.

Now, let us focus on the peak position $x(t)$ of the transmitted packet and compare it with the peak position $x_{\mathrm{f}}(t)$ of the free wave packet that would have evolved under the same initial condition without the barrier. Arguing that the spatial shift of the peak, $\Delta x(t) \equiv x(t) - x_{\mathrm{f}}(t)$, is due to the extra time spent by the tunneling particle inside the barrier, we are led to the following expression for the tunneling time, which is called the *phase time*,

$$\tau_{\text{phase}} = \tau_{\text{free}} + \Delta x(t)/v_0\,, \tag{6.9}$$

where $\tau_{\text{free}} \equiv d/v_0$ is the time a classical free particle with velocity v_0 spends in the spatial region $a < x < b$. Now, the transmitted packet is expressed as

$$\int \frac{\mathrm{d}k}{\sqrt{2\pi}}\, \psi(k-k_0) T(k) \mathrm{e}^{\mathrm{i}kx - \mathrm{i}\hbar k^2 t/2m}\,. \tag{6.10}$$

times. In text books, the tunneling probability for the stationary case is usually defined as the ratio of transmitted flux to incident flux (with a velocity ratio factor when the barrier is not symmetrical). This procedure is, however, outside the standard prescription for calculating probabilities using amplitude and operators. The procedure presented here is completely within the standard prescription.

Because of the peaked nature of $\psi(k - k_0)$, we can evaluate (6.10) using the stationary phase approximation. We then find that the peak position of the transmitted packet moves as $x(t) = v_0 t - \theta'(k)|_{k=k_0}$, where θ is the phase defined by (6.5) and the prime denotes $\partial/\partial k$, while the peak of the free wave packet moves as $x_{\mathrm{f}}(t) = v_0 t$. Using these results in (6.9), we have

$$\tau_{\text{phase}} = \hbar \left. \frac{\partial \phi}{\partial E} \right|_{E=E_0} , \tag{6.11}$$

where $E = \hbar^2 k^2/2m$, $E_0 \equiv \hbar^2 k_0^2/2m$, and ϕ is defined by (6.5). Essentially the same result can be obtained by following, instead of the peak of the transmitted packet, a suitably defined 'center of gravity'.

The phase time is undoubtedly useful in a qualitative understanding of the motion of the transmitted packet. However, we cannot consider it to be the time taken by a packet peak to traverse the barrier because the peak of the transmitted packet is not clear near the barrier. In the above derivation, the clear peak motion in the region $x \gg b$ was linearly extrapolated back to $x = b$. Near the barrier, the extrapolated peak motion, which is the basis of (6.11), cannot be found in the real dynamics of the wave function. The phase time is useful but is not the answer to the tunneling time problem.

6.2.3 Velocity Under the Barrier

The phase time method is based on the evaluation of (6.10) by the stationary phase approximation. What about more sophisticated evaluations to access the motion inside the barrier? Using a contour integral representation of the time dependent wave function, Stevens [18] analyzed how a pulse wave that initially has a sharp wave front (product of a plane wave and a step function) propagates under the step potential region of height V_0. Evaluating the integral by the steepest descent method, he concluded that the wave front propagates at a velocity $v_E \equiv \sqrt{2(V_0 - E)/m}$.[7] This conclusion had an impact because a clear velocity was assigned to the motion of the particle in the classically forbidden region. However, Teranishi et al. [19] reinvestigated the problem with more care in the analytical steps and found that Stevens' conclusion is incorrect. For the case of a square barrier, this negative result was also confirmed numerically by Jauho and Jonson [20] and analytically by Ranfagni et al. [21]. Moretti [22] arrived at a conclusion similar to Stevens for a square barrier, but the result attributing a velocity to the particle under the barrier was questioned by Muga et al. [23].

Although Stevens's original result is not correct, it motivated others to develop an exact analytical expression[8] for the time-dependent wave function. The present author considers this exact solution approach to be very

[7] This implies that the tunneling time for an opaque rectangular barrier is d/v_E, which is identical to the Büttiker–Landauer time to be discussed in Sect. 6.2.5.

[8] In terms of the critical points (such as poles) of the transmission and reflection amplitudes.

important because it allows a non-perturbative analysis of tunneling in an analytical manner.[9] For this approach, see [23–26]. Of course, if one aims simply to visualize the time evolution of the wave function, one can numerically solve the Schrödinger equation from the beginning. There are many papers on this issue including the very early one by Goldberg et al. [27].

6.2.4 The Larmor Clock

It may be said that the phase time is based on an equation of type $x = vt$ in real space, while the Larmor times discussed here are based on an analogous equation of type $\varphi = \omega t$ in spin space with φ being a spin rotation angle and ω an angular velocity. There is, however, an important difference between the two. That is, the phase time method is based on $x = vt$ outside the barrier, while the Larmor clock method is based on $\varphi = \omega t$ inside the barrier. In the phase time method, x and v (to be precise, Δx and v_0), and therefore the resultant t, were quantities associated with the motion of wave packets outside the barrier. If we attempt to use an equation of type $x = vt$ inside the barrier, which is preferable since we are interested in the time inside the barrier, we immediately face the difficulty that v is not meaningful (or it is imaginary) in the barrier. This kind of difficulty is absent in the Larmor clock method.

We consider the tunneling of a particle with spin 1/2. The motion of the particle is supposed to be one-dimensional along the y-axis. A small and homogeneous magnetic field $B_0\boldsymbol{z}$ is applied only in the barrier region ($\boldsymbol{z}$ is the unit vector in the z-direction). The Hamiltonian now has an additional term

$$-\frac{\hbar\omega_{\mathrm{L}}}{2}\sigma_z\Theta_{ab}(y)$$

on the right-hand side of (6.1) (replac x by y). Here σ_z is a Pauli spin matrix. The initial spin is supposed to be polarized in the x-direction, and the initial

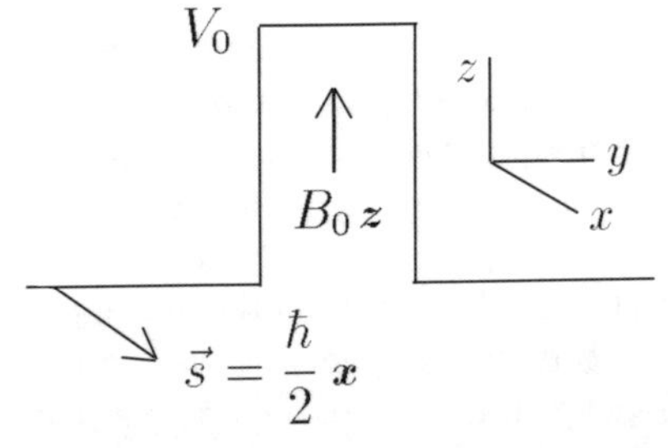

Fig. 6.4. The Larmor clock method

[9] Of course, the critical points must be obtained numerically.

orbital wave function is assumed to be monochromatic with wave number k_0 (energy E_0).

As the particle enters the barrier, the spin starts to rotate due to the magnetic field. If the particle is at rest in the magnetic field, the spin rotates in the xy-plane through an angle $\omega_{\rm L} \equiv g\mu_{\rm B}B_0/\hbar$ per unit time, where g is the gyromagnetic ratio, and $\mu_{\rm B}$ the Bohr magneton. This is a Larmor precession. Hence the rotation angle divided by $\omega_{\rm L}$ gives the time the particle is located in the magnetic field. In the tunneling case, it thus also seems natural to calculate the tunneling time by dividing the spin rotation angle accumulated in the barrier by $\omega_{\rm L}$. However, as pointed out by Büttiker [28], the motion of the spin is no longer a Larmor precession in the xy-plane in the tunneling case. Indeed, the spin rotates not only in the xy-plane but also in the xz-plane. Only the rotation in the xy-plane was considered in earlier papers [29,30]. Therefore, we obtain two different tunneling times depending on which rotation is analyzed with an equation of type $\varphi = \omega t$.

That two rotations arise can be understood as follows. The initial spin may be written as the superposition of a spin-up and a spin-down state ('up' and 'down' are with respect to the z-direction). Due to Zeeman splitting, the spin-up component sees a modified barrier of height $V_0 - \hbar\omega_{\rm L}/2$, while the spin-down component sees a modified barrier of height $V_0 + \hbar\omega_{\rm L}/2$. Therefore, the spin-up component is preferably transmitted since the barrier is lower, and as a result the spin of the transmitted particle acquires a non-zero z component. Of course, the spin also acquires a y component due to the conventional Larmor precession. The expectation values of s_z and s_y for transmitted particles can be written in terms of the transmission amplitudes for the square barriers of height $V_0 \pm \hbar\omega_{\rm L}/2$. To the lowest order in B_0, they are found to be

$$\langle s_z \rangle = \frac{\hbar}{2}\omega_{\rm L}\tau_z, \quad \tau_z = -\hbar\frac{\partial \ln|T|}{\partial V_0} \approx \frac{md}{\hbar\kappa}, \tag{6.12}$$

$$\langle s_y \rangle = -\frac{\hbar}{2}\omega_{\rm L}\tau_y, \quad \tau_y = -\hbar\frac{\partial \phi}{\partial V_0} \approx \frac{\hbar k_0}{V_0\kappa}, \tag{6.13}$$

where $\kappa \equiv \sqrt{k_{V_0}^2 - k_0^2}$, and the last expressions in (6.12) and (6.13) are for opaque barriers. We thus end up with the two tunneling times τ_z and τ_y. One can of course take an average of the two tunneling times to define a single quantity such as $\tau_{\rm T} \equiv \sqrt{\tau_z^2 + \tau_y^2}$, which is Büttiker's *traversal time* (the subscript T in $\tau_{\rm T}$ stands for 'traversal'). However, there are many ways to define an average and, at least within the Larmor clock approach, a special status cannot be given to $\tau_{\rm T}$. The Larmor clock method does not give a unique answer to the tunneling time problem.

In a narrower sense, τ_y is called the Larmor time, but in a broader sense both τ_y and τ_z are often called the Larmor times. Leavens call τ_y the 'spin-precession traversal time of Rybachenko' and τ_z the 'spin-rotation traversal time of Büttiker'.

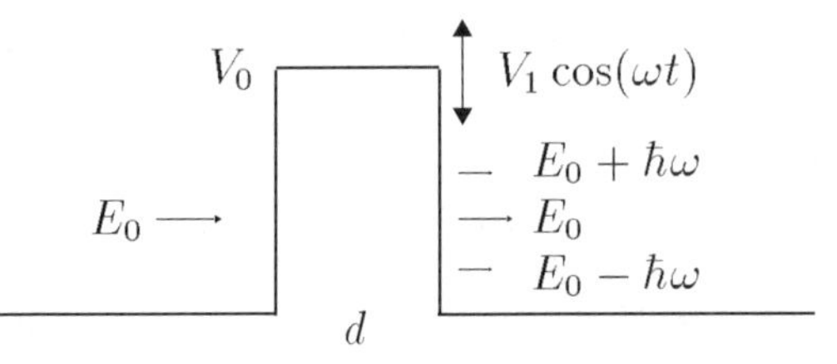

Fig. 6.5. Oscillating barrier of Büttiker and Landauer

Leavens [31] generalized Büttiker's analysis of the Larmor clock to the case of arbitrary barriers by introducing an average height $\bar{V}$ of the potential, finding that the resultant Larmor times are also given by (6.12) and (6.13) with V_0 replaced by $\bar{V}$. Leavens [32] also made a generalization to the case of a uniform transverse magnetic field confined to an arbitrary part of an arbitrary barrier. Important findings are:

- each of the resultant four Larmor times (τ_z and τ_y for reflection and transmission) is spatially additive, i.e., each satisfies (6.30) for $0 < x_1 < x_2 < x_3 < d$,
- the reflection and transmission times are consistent with the local dwell time. For example, τ_z for reflection and τ_z for transmission satisfy (6.29).

A series of precise measurements of Larmor times has been reported by a Japanese group. The results are, as they must be, in excellent agreement with theoretical predictions. See [33] and references therein.

6.2.5 Time-Modulated Barriers

The previous two approaches attempt to measure the tunneling time directly. The peak shift or spin rotation angles are measured and translated into time information through equations of the form $x = vt$ or $\varphi = \omega t$. In contrast, the approach to be discussed here attempts to measure the tunneling time indirectly. The tunneling particle is perturbed with a known time scale and the tunneling time is then estimated by finding such a time scale of perturbation that leads to apparent changes in the behavior of the tunneling particle.

Büttiker and Landauer [34,35] studied an oscillatory barrier, where a small perturbation $V_1 \cos(\omega t)$ is added to the original static barrier. The Hamiltonian is therefore given by (6.1) with V_0 replaced by $V_0 + V_1 \cos(\omega t)$. Since the potential is time dependent, the energy of the transmitted particle can differ from that of the incident particle, which is assumed to be E_0 (monochromatic). The particle can emit or absorb energy quanta $n\hbar\omega$ (n being a positive integer) through the interaction with the barrier. Treating V_1 as a perturbation and assuming further that $\hbar\omega \ll E_0$, $\hbar\omega \ll V_0 - E_0$, and $d\sqrt{2m(V_0 - E_0)}/\hbar \gg 1$, they analyzed the case of emission or absorption of a single quantum $\hbar\omega$ and calculated the tunneling probabilities P_+ and P_-

that the particle is transmitted at the sideband energies $E_0 + \hbar\omega$ and $E_0 - \hbar\omega$, respectively. They showed for an opague barrier, with $T = 2\pi/\omega$,

$$P_\pm = \left(\frac{V_1}{2\hbar\omega}\right)^2 \left[\exp(\pm 2\pi\tau_{\rm BL}/T) - 1\right]^2 P\,, \tag{6.14}$$

$$\tau_{\rm BL} \equiv \frac{md}{\hbar\kappa} = \frac{d}{v_E}\,, \quad v_{\rm E} \equiv \sqrt{2(V_0 - E_0)/m}\,, \tag{6.15}$$

where P is the tunneling probability at the energy E_0, T is the modulation period (not the transmission amplitude), and $v_{\rm E}$ is the absolute value of the imaginary velocity $\sqrt{2(E_0 - V_0)/m}$ under the barrier (here the subscript E on $v_{\rm E}$ stands for 'Euclidean').

Let us consider $P_\pm$ as functions of the modulation period T and observe how the functions behave as T is varied. Equation (6.14) shows that $P_\pm/P$ is negligible for $2\pi\tau_{\rm BL}/T \ll 1$, while it is not for $2\pi\tau_{\rm BL}/T > 1$. Because of this crossover behavior at $T \sim 2\pi\tau_{\rm BL}$, we identify $\tau_{\rm BL}$ as the tunneling time, apart from the numerical factor 2π. For opaque barriers, we can write

$$\tau_{\rm BL} \approx -\hbar \frac{\partial \ln |T|}{\partial V_0}\,. \tag{6.16}$$

Normally, $\tau_{\rm BL}$ is called the Büttiker–Landauer time, although Hauge and Støvneng [2] called $\tau_{\rm T} = \sqrt{\tau_z^2 + \tau_y^2}$ the Büttiker–Landauer time. For opaque barriers, $\tau_{\rm T} \approx \tau_{\rm BL}$ since $\tau_z = \tau_{\rm BL} \gg \tau_y$.

As just mentioned, $\tau_{\rm BL}$ coincides with τ_z obtained in the Larmor clock method. Under the name of 'bounce time', $\tau_{\rm BL}$ also appears in the instanton technique, an imaginary-time path integral technique for calculating tunneling probabilities. Does this popularity of $\tau_{\rm BL}$ imply that it is *the* tunneling time?

Before giving $\tau_{\rm BL}$ a special status, it must be checked whether a different modulation of the barrier also gives $\tau_{\rm BL}$ as a time scale characterizing some crossover behavior. To this end, Takagi [36] studied another time-modulated barrier model, where not only the height but also the width of the barrier changes in time (see Fig. 6.6). In particular, the tunneling through a *squeezing potential* was studied. The height $V(t)$ and width $d(t)$ change as

$$V(t) = \frac{V_0}{(1 + t/T)^2}\,, \quad d(t) = (1 + t/T)d\,. \tag{6.17}$$

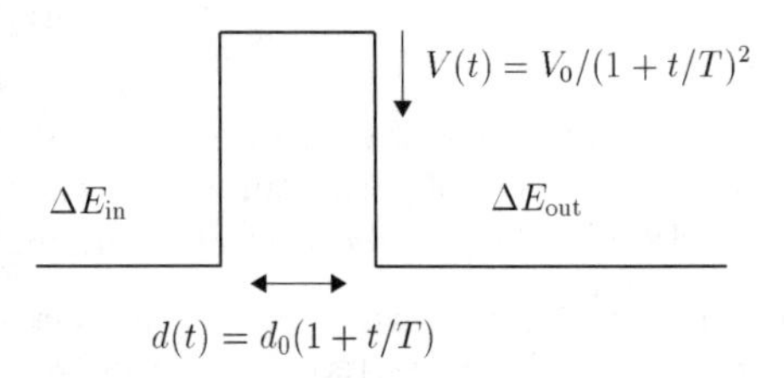

Fig. 6.6. Linearly squeezing barrier of Takagi

A remarkable feature of this model is that the exact expression for an evolving wave packet can be obtained in terms of the scattering amplitudes. Since the potential changes with time scale T, the energy of the transmitted particle can differ from the incident energy. This time, however, the energy does not shift by an amount $n2\pi\hbar/T$ because the barrier modulation is not periodic. Thus, instead of $P_\pm/P$, we need a new measure that quantifies how the tunneling is affected by the barrier modulation. Paying attention to the barrier effect on the transmitted wave function in k space, rather than real space, Takagi used $\alpha \equiv \Delta p_{\rm out}/\Delta p_{\rm in} - 1$ as the measure, where $\Delta p_{\rm in}$ is the momentum uncertainty of the incident packet and $\Delta p_{\rm out}$ is that of the transmitted packet. Analyzing α as a function of T, he showed that for opaque barriers α is negligible for $\tau_{\rm Takagi}/T \ll 1$, while it is not for $\tau_{\rm Takagi}/T > 1$, where

$$\tau_{\rm Takagi} \equiv \frac{2m}{\hbar} \left.\frac{\partial^2 \phi}{\partial k^2}\right|_{k=k_0} \approx \frac{\hbar^2 k_0}{m^2 v_{\rm E}^3} . \tag{6.18}$$

If we follow the spirit of Büttiker and Landauer, $\tau_{\rm Takagi}$ should now be identified as the tunneling time.

Depending on how we modulate the barrier, we now get different answers.[10] The difference between $\tau_{\rm BL}$ and $\tau_{\rm Takagi}$ is not a minor difference, either qualitatively or quantitatively. Like the Larmor clock approach, the modulated barrier approach also fails to give a unique answer to the tunneling time problem.

Note that the status of $1/\tau_{BL}$ as the crossover frequency in the oscillating barrier model has been questioned by some authors [38–40]. They numerically calculated the transmission probability as the function of ω but failed to find a crossover behavior at $\omega \sim 1/\tau_{BL}$.

Although $\tau_{\rm BL}$ is widely used as *the* tunneling time in the literature, its status must be considered as not yet established.

6.2.6 Dwell Time and its Decomposition

Applying the concept of collision time [41] in the scattering theory to the case of the one-dimensional stationary scattering problem, Büttiker [28] introduced the dwell time $\tau_{\rm d}$ as the ratio of the number of particles within the barrier to the incident flux:

$$\tau_{\rm d}(k) = \frac{m}{\hbar k} \int_a^b {\rm d}x \, |u_k(x)|^2 , \tag{6.19}$$

where $u_k(x)$ is the eigenfunction with wave number k. As clearly stated by Landauer and Martin [1], the meaning of $\tau_{\rm d}$ is simply 'the time that the incident flux has to be turned on to provide the accumulated particle storage

[10] This is also supported by the study due to Truscott [37], who analyzed another time-modulated barrier but did not find $\tau_{\rm BL}$ as the characteristic time in the sense of Büttiker and Landauer.

in the barrier'. In the literature (including [28]), τ_d is often called the 'average' dwell time. However, since it is not the case that τ_d is given as an average with respect to a probability distribution (of dwell time), it is not appropriate to regard it as an averaged quantity.[11] It is widely believed that (6.19) is the (average) time spent by a particle with wave number k in the barrier region irrespective of whether it is finally reflected or transmitted.[12]

The issue of dwell time is one of the major sources of controversy. A question arises when one tries to generalize (6.19) to the case of a wave packet. Leavens and Aers [42] pointed out that an interesting relation

$$\int_{-\infty}^{\infty} \mathrm{d}t \int_{a}^{b} \mathrm{d}x \, |\Psi(x,t)|^2 = \int_{-\infty}^{\infty} \frac{\mathrm{d}k}{2\pi} \, |\psi(k)|^2 \, \tau_d(k) \tag{6.20}$$

follows from the results of Hauge, Falck, and Fjeldly [43], where $\psi(k)$ is the k-space wave function obtained from $\Psi(x,t)$ via Fourier transform. The right-hand side of (6.20) is a clear expression as an average. Indeed, it is the k average of $\tau_d(k)$ with the k-space probability distribution $|\psi(k)|^2$. Leavens and Aers [42] seem to have considered that the meaning of the space–time integral [the left hand side of (6.20)] was undoubtedly an exact expression for the average time spent in the barrier region $a \leq x \leq b$ by a particle with wave function $\Psi(x,t)$. They then read (6.20) from left to right to justify the interpretation of $\tau_d(k)$ as the dwell time for a given $k > 0$ ($|\psi(k)|^2 \approx 0$ was assumed for $k < 0$). However, as pointed out by Hauge and Støvneng (Appendix B of [2]), who although are in favor of the above interpretation of the space–time integral, the meaning of the space–time integral is not self-evident in conventional quantum mechanics.

Despite this subtlety, the space–time integral is now widely believed to be the correct expression for the average dwell time when the particle is described by a wave packet. This is probably because the space–time integral has been reached in several different ways [the range of the time integral is not necessarily $(-\infty, \infty)$] in the analysis of tunneling time or related matters. Baskin and Sokolovski [44] and Sokolovski and Baskin [45] obtained the space–time integral from a Feynman paths average of a traversal time operator that is non-local in time and acts on Feynman trajectories. Leavens and Aers [46] showed that, within Bohmian mechanics, the space–time integral has a clear meaning as an average dwell time. Jaworski and Wardlaw [47,48] also postulated the space–time integral as giving the mean sojourn

[11] The word 'average' (or 'mean') is widely abused in the literature. We often encounter phrases such as 'average' dwell time, 'average' transmission time, and 'average' reflection time without finding the relevant probability distributions. The word 'average' should be used only when the relevant probability distribution exists. Otherwise, one should write, for example, 'a measure of dwell time' instead of 'the average dwell time'.

[12] It seems to the present author that by taking this for granted one goes beyond the original meaning of $\tau_d(k)$ so clearly stated by Landauer.

(dwell) time in one-dimensional scattering. At the same time, they well recognized that the origin and interpretation of the space–time integral is not self-evident because of the absence of real physical trajectories in quantum mechanics [48].

Is the space–time integral indeed a general expression for the average dwell time for the particle described by a wave packet? Or is it just a beautiful way of expressing the right-hand side of (6.20)? The right-hand side is indeed the k average of $\tau_{\mathrm{d}}(k)$ but it is not a quantum mechanical average since it has not been derived from a general prescription for calculating an average from amplitudes and operators. The present author considers that, in order to interpret the space–time integral as an average dwell time, one must show the underlying probability distribution of dwell time within conventional quantum mechanics. From this viewpoint, it is safe to say that the status of the space–time integral is still not clear. Apart from these conceptual questions, the dwell time for stationary states is widely used in the literature, especially in the analysis of transport properties of semiconductor heterostructures. See, for example, [49].

Another question that has attracted people is whether or not one can decompose the dwell time into transmission and reflection components as

$$\tau_{\mathrm{d}} = P_{\mathrm{t}}\tau_{\mathrm{t}} + P_{\mathrm{r}}\tau_{\mathrm{r}}\,, \tag{6.21}$$

in which the dwell time is supposed to be given by (6.19) or (6.20), τ_{t} is the transmission time, the time a transmitted particle spends in the barrier region, and τ_{r} is the reflection time, the time a reflected particle spends in the barrier. P_{t} is the transmission probability and P_{r} the reflection probability ($P_{\mathrm{t}} + P_{\mathrm{r}} = 1$). When discussing the above decomposition, we give a special status to the dwell time. That is, we simply accept (or assume) that the left-hand side of (6.21) is given by (6.19) or (6.20) with the interpretation that it is the average dwell time for a particle in a stationary state or in a localized state described by a wave packet. This was criticized by Olkhovsky and Recami [3]. Landauer and Martin [1] and Büttiker [4] also criticized (6.21) as a meaningless relation in quantum mechanics. On the other hand, Hauge and Støvneng [2] took the 'probabilistic identity' (6.21) for granted and used it to single out the correct transmission and reflection times from various times that had been proposed.

In Sect. 6.2.2, we considered the transmission time τ_{phase} by following the peak motion of a transmitted packet. We can also follow the peak of the reflected packet to obtain the corresponding reflection time. For the time being let us denote these transmission and reflection times by τ_{t}^{ϕ} and τ_{r}^{ϕ}, respectively. Hauge, Falck, and Fjeldly [43] found that the transmission phase time τ_{t}^{ϕ} and the reflection phase time τ_{r}^{ϕ} satisfy the following equation that has an interference term on the right-hand side:

$$\tau_{\mathrm{d}}(k) = P_{\mathrm{t}}\tau_{\mathrm{t}}^{\phi} + P_{\mathrm{r}}\tau_{\mathrm{r}}^{\phi} + \frac{m}{\hbar k^2}\sqrt{P_{\mathrm{r}}}\sin(\beta - 2ka)\,, \tag{6.22}$$

where $P_t = |T|^2$, $P_r = |R|^2$, and β is defined by $R = |R|e^{i\beta}$ [see (6.3)]. This relation was generalized in [2] to the case of a wider region of space including the barrier region, to stress the meaningfulness of asymptotic phase times when they are properly averaged.

Below, we review other approaches from the viewpoint of whether they yield transmission and reflection times that satisfy (6.21).

6.2.7 Feynman Paths Approach

If a particle follows a well-defined path in space–time, the tunneling time problem disappears. But there is no such real physical path in quantum mechanics. Instead, quantum mechanics can be formulated with various types of quantum trajectories: Feynman paths, Bohm trajectories, Nelson trajectories, and Wigner trajectories. The first three are trajectories in space–time, while the last one is defined in phase space. It is natural to attempt to use these trajectories to define and analyze the tunneling time. Here we examine a Feynman path approach.[13]

Consider Feynman paths connecting given end points (x_1, t_1) and (x_2, t_2) $(t_1 < t_2)$. We start by considering the sum over those Feynman paths that spend a certain amount of time τ in the barrier region:

$$F_{2,1}(\tau) \equiv \frac{\sum_{\text{paths}} \delta(\tau_{ab}[x(\cdot)] - \tau) e^{iS[x(\cdot)]/\hbar}}{\sum_{\text{paths}} e^{iS[x(\cdot)]/\hbar}} , \tag{6.23}$$

where $\sum_{\text{paths}}$ stands for a sum over paths, the denominator (introduced to normalize $F_{2,1}$) is the sum over all paths and is therefore the propagator between the end points, $\tau_{ab}[x(\cdot)]$ is the amount of time a Feynman path $x(t)$ spends in the barrier region [for its precise definition, see (6.53)], and the initial point x_1 is assumed to be on the left side of the barrier, i.e., $x_1 < a$.

Now, $F_{2,1}(\tau)$ is a distribution of τ at the amplitude level (a complex-valued 'amplitude distribution'). As pointed out by Landauer and Martin [1], it is not clear what procedure is needed to extract physical information about the time under the barrier from such an amplitude distribution, which is not the usual wave function. Nevertheless, since $\int d\tau F(\tau) = 1$, let us formally regard $F_{2,1}(\tau)$ as a 'probability' distribution for τ and calculate its 'average' according to $\int d\tau\, \tau F(\tau)$. We expect to obtain a transmission time for $x_2 > b$ and a reflection time for $x_2 < a$, which we shall denote by τ_1 and τ_2, respectively. Sokolovski and Baskin [45] (see also [44]) found

$$\tau_1(k) = i\hbar \int_a^b dx \frac{\delta \ln T}{\delta V(x)} = i\hbar \left. \frac{\partial \ln T(V_0 + V, k)}{\partial V} \right|_{V=0} , \tag{6.24}$$

[13] The approach presented here is mainly due to Sokolovski and coworkers and also to Fertig. For Feynman paths approaches by other authors, see [50–52]. For the Nelson trajectory approach, see [53–55]. For the Wigner trajectory approach, see [56,57]. The Bohm trajectory approach by Leavens is discussed in Sect. 6.2.10. A Bohm trajectory approach by other authors is found in [58].

and a similar expression for $\tau_2(k)$ (just replace T above by R), where V_0 is the height of the potential under consideration.[14] A remarkable feature of these times is that they satisfy [44,45,88]

$$\tau_{\mathrm{d}}(k) = P_{\mathrm{t}}\tau_1(k) + P_{\mathrm{r}}\tau_2(k)\,. \tag{6.25}$$

However, this is not the desired decomposition because $\tau_1(k)$ and $\tau_2(k)$ are complex-valued. Note that Sokolovski and Baskin's complex tunneling times are written in terms of the potential height derivative of the transmission amplitude. Earlier, Pollak and Miller [59] extracted another complex interaction time from the flux–flux correlation function, a tool widely used in chemical physics to calculate rate constants and reaction rate probabilities in chemical reactions. The resultant complex time is similar to Sokolovski and Baskin's but the derivative is with respect to the energy of the particle rather than the potential height. Martin and Landauer [60] also derived complex times written in terms of energy derivatives of the transmission amplitude.

Baskin and Sokolovski also found a close relationship between their complex times and the Larmor times [44,45]:

$$\tau_1(k) = \tau_{\mathrm{t}y}(k) - \mathrm{i}\tau_{\mathrm{t}z}(k)\,, \quad \tau_2(k) = \tau_{\mathrm{r}y}(k) - \mathrm{i}\tau_{\mathrm{r}z}(k)\,, \tag{6.26}$$

where $\tau_{\mathrm{t}y}(k)$ and $\tau_{\mathrm{t}z}(k)$ are, respectively, τ_y and τ_z in (6.13) and (6.12), and $\tau_{\mathrm{r}z}(k)$ and $\tau_{\mathrm{r}y}(k)$ are the Larmor times for reflection [28]. Since the left-hand side of (6.25) is positive definite, taking the real part of (6.24) gives an equation of the form (6.21) with real-valued transmission and reflection times. We can also take the imaginary part. Equation (6.25) can then be split into[15]

$$\tau_{\mathrm{d}}(k) = P_{\mathrm{t}}\tau_{\mathrm{t}y}(k) + P_{\mathrm{r}}\tau_{\mathrm{r}y}(k)\,, \tag{6.27}$$

$$0 = P_{\mathrm{t}}\tau_{\mathrm{t}z}(k) + P_{\mathrm{r}}\tau_{\mathrm{r}z}(k)\,. \tag{6.28}$$

Now, both $\tau_{\mathrm{t}y}$ and $\tau_{\mathrm{r}y}$ are real. Unfortunately, however, $\tau_{\mathrm{r}y}$ is not positive definite and therefore (6.27) is not the desired decomposition.[16] There are still other reasons to doubt the status of $\tau_{\mathrm{t}y}$ and $\tau_{\mathrm{r}y}$ as physically reliable transmission and reflection times.

[14] Note that $\tau_1(k)$ and $\tau_2(k)$ do not depend on the end points. Sokolovski and Baskin [45] derived an approximate expression for the propagator in the presence of a barrier. Using this expression, it is easy to see that the denominator and the numerator on the right-hand side of (6.23) depend on the end points in the same way. Thus the final results are independent of the end points.

[15] For a rectangular barrier, $\tau_{\mathrm{t}y} = \tau_{\mathrm{r}y}$ because T and R have the same phase factor (apart from a constant). Using this property (and also $P_{\mathrm{t}} + P_{\mathrm{r}} = 1$) in (6.27), we obtain $\tau_{\mathrm{d}}(k) = \tau_{\mathrm{t}y} = \tau_{\mathrm{r}y}$ for a rectangular barrier. The second equation (6.28) expresses conservation of angular momentum. These were pointed out in [28].

[16] It should be noted, however, that the positivity requirement on tunneling times is somewhat subtle. See, for example, Sect. 1 of [5] and also p. 806 of [61].

- Leavens and Aers (Sect. IV of [62]) and Hauge and Støvneng (Sect. VI C3 of [2]) presented an example where taking τ_{ty} as a physically reliable transmission time betrays our intuition that 'a reflected particle spends exactly zero time on the far side of a barrier' [62].
- For an opaque barrier, τ_{ty} is almost independent of the barrier width. Thus, for a sufficiently thick barrier, the speed v_t under the barrier calculated from the classical equation $v_t = d/\tau_{ty}$ can be greater than the speed of light. Leavens and Aers [62,63] showed that this is also the case if the Dirac equation is used instead of the Schrödinger equation, thus expressing a serious doubt about the status of τ_{ty} as a physically meaningful transmission time [5,62].

These arguments remind us of the status of (6.21), i.e., that it is only a necessary condition for tunneling times even if it is correct. Leavens and Aers [5] proposed the following set of conditions to be satisfied by physically meaningful tunneling times (for a stationary case):

(i) (6.21) or (ii) its generalization

$$\tau_{\rm d}(k;x_1,x_2) = P_{\rm t}\tau_{\rm t}(k;x_1,x_2) + P_{\rm r}\tau_{\rm r}(k;x_1,x_2)\,, \tag{6.29}$$

where $P_{\rm t} = |T(k)|^2$, $P_{\rm r} = |R(k)|^2$, and $\tau_i(k;x_1,x_2)$ is the dwell time (i = d), transmission time (i = t), and reflection time (i = r) for the interval $x_1 < x < x_2$,

(iii) additivity

$$\tau_i(x_1,x_3) = \tau_i(x_1,x_2) + \tau_i(x_2,x_3) \quad \text{for} \quad x_1 < x_2 < x_3\,, \tag{6.30}$$

(iv) positivity

$$\tau_i(x_1,x_3) > 0\,, \tag{6.31}$$

(v) consistency with causality for $\tau_{\rm t}$.

Sokolovski and coworkers made an extensive contribution to the tunneling time problem by using Feynman paths. Readers are also referred to [64–68] for their work, and also to [69], in which the Feynman paths approach and the Bohm trajectory approach to be discussed later are reviewed and compared.

6.2.8 Systematic Projector Approach

Brouard, Sala, and Muga [70] (see also [71,72]) have developed a systematic method for decomposing the 'mean' dwell time [postulated to be given by the space–time integral in (6.20)] into transmission and reflection times. This *systematic projector approach* yields an infinite number of resolutions of dwell time, giving, in general, equations of the form

$$\tau_{\rm d} = P_{\rm t}\tau_{\rm t} + P_{\rm r}\tau_{\rm r} + \text{interference term}\,. \tag{6.32}$$

The systematic projector approach contains three projection operators P, $Q(= 1 - P)$, where 1 is the identity operator, and D. The projector P selects that part of $\Psi(x,t)$ associated with transmission (in the sense that the selected part has only the positive momentum components in the infinite future), and the projector Q selects the part associated with reflection (only negative momentum components in the infinite future). The expectation values of P and Q are, respectively, the usual transmission and reflection probabilities defined in the limit $t \to \infty$ and hence independent of time. The projector D selects the part of $\Psi(x,t)$ inside the barrier region $a < x < b$ so that its expectation value $\langle D\rangle \equiv \langle\Psi(t)|D|\Psi(t)\rangle$ is the usual probability of finding the particle in the barrier region at time t. The left-hand side of (6.20) can thus be written as $\int\langle D\rangle\,\mathrm{d}t$. The essence of the systematic projector approach is to rewrite D in $\int\langle D\rangle\,\mathrm{d}t$ using the property $P + Q = 1$ and the idempotency of a projection operator (e.g., $D = DD$). For instance, we can write $D = PD+QD$, $D = DP+DQ$, and $D = PDP+QDQ+PDQ+QDP$. Using the first one in $\int\langle D\rangle\,\mathrm{d}t$ leads to [71]

$$\tau_{\mathrm{d}} = P_{\mathrm{t}}\tau_{\mathrm{t}}^{PD} + P_{\mathrm{r}}\tau_{\mathrm{r}}^{QD}\,, \tag{6.33}$$

where

$$\tau_{t}^{PD} \equiv \frac{1}{P_{\mathrm{t}}}\int\langle\Psi(t)|PD|\Psi(t)\rangle\,\mathrm{d}t,\ \tau_{\mathrm{r}}^{QD} \equiv \frac{1}{P_{\mathrm{r}}}\int\langle\Psi(t)|QD|\Psi(t)\rangle\,\mathrm{d}t\,. \tag{6.34}$$

Equation (6.33) does not have an interference term, but τ_{t}^{PD} and τ_{r}^{QD} are complex-valued since PD and QD are not Hermitian. Muga, Brouard and Sala [71] found that the two complex times are nothing but Sokolovski and Baskin's complex-valued transmission and reflection times, i.e., $\tau_{\mathrm{t}}^{PD} = \tau_1$ and $\tau_{\mathrm{r}}^{QD} = \tau_2$ [see (6.26) for τ_1 and τ_2].

Now, the second choice $D = DP+DQ$ leads us to another decomposition similar to (6.33) with new transmission and reflection times, which we denote by τ_{t}^{DP} and τ_{t}^{DQ}, respectively. Since D and P do not commute, $\tau_{\mathrm{t}}^{PD} \neq \tau_{\mathrm{t}}^{DP}$ and $\tau_{\mathrm{r}}^{QD} \neq \tau_{\mathrm{r}}^{DQ}$, and therefore different decompositions of D lead to different expressions for τ_{t} and τ_{r} in general. The third choice [71]

$$D = PDP + QDQ + PDQ + QDP$$

leads us to yet another decomposition of the form of (6.32) with the interference term given by $2\mathrm{Re}[\tau_{\mathrm{int}}]$, where

$$\tau_{\mathrm{int}} \equiv \int\langle\Psi(t)|PDQ|\Psi(t)\rangle\,\mathrm{d}t\,, \tag{6.35}$$

which might be called an interference time. The real-valued times τ_{t}^{PDP}, τ_{r}^{QDQ}, and τ_{int} obtained from the present decomposition coincide with the transmission, reflection, and interference times introduced by van Tiggelen, Tip, and Lagendijk [73]. In this way, the systematic projector approach reproduces many of the proposed transmission and reflection times in a systematic way. Leavens [61,74] has shown, however, that the systematic projector approach does not give positive definite transmission and reflection times that satisfy both the probabilistic identity (6.21) and the condition for additivity (6.30). The utility of the systematic approach is that it provides, completely within standard quantum mechanics, 'a compact classification scheme for many of the proposed times and simplifies their connection and study' [23].

In the systematic projector approach, quantities arise naturally that we are tempted to interpret as joint probabilities. In the third choice discussed above, we encounter a quantity $\langle\Psi(t)|PDP|\Psi(t)\rangle$. This has a suggestive form as the joint probability that 'the particle is now transmitted *and* was in the barrier region before. Similarly, $\langle\Psi(t)|DPD|\Psi(t)\rangle$ is a suggestive form as the joint probability that 'the particle is now in the barrier region *and* will eventually be transmitted'. The two joint probabilities must coincide in classical probability theory, but they do not in fact because D and P do not commute. This simply means that the concept of joint probability cannot be introduced in quantum mechanics when the relevant observables do not commute. In a word, the 'sample space' for the desired joint probabilities is not an exhaustive set of mutually exclusive events. Rather, the 'sample space' consists of interfering alternatives, for which probabilities are not definable in a consistent way as in the case of the two-slit experiment. Even so, the use of such 'generalized probabilities' which are not actual probabilities is still interesting if one can extract useful information and/or if it throws new light on known results.

6.2.9 Weak Measurement Approach

Steinberg [75,76] and Iannaccone [77] have shown that the complex-valued conditional probabilities motivated by the weak measurement theory of Aharonov, Albert, and Vaidman [78,79] yield some of the known results about tunneling time. Note that (6.21) is immediate from

$$|\Psi(x,t)|^2 = P_{\mathrm{t}} P(x,t|\mathrm{trans}) + P_{\mathrm{r}} P(x,t|\mathrm{refl})\,, \tag{6.36}$$

provided that the conditional probabilities $P(x,t|\mathrm{trans})$ and $P(x,t|\mathrm{refl})$ are definable, where the former is supposed to be the probability distribution of the particle's position at time t conditioned on the information that the particle is found to have been transmitted at sufficiently later times, and similarly for $P(x,t|\mathrm{refl})$. An equation of the form of (6.36) is immediate from the identity

$$\langle i|A|i\rangle = \sum_n |\langle n|i\rangle|^2 A_{\mathrm{w}}^{(n)}\,, \quad \text{where} \quad A_{\mathrm{w}}^{(n)} \equiv \frac{\langle n|A|i\rangle}{\langle n|i\rangle}\,. \tag{6.37}$$

Let us put, with $\hbar = 1$,

$$|i\rangle = |\Psi(t)\rangle \equiv \mathrm{e}^{-\mathrm{i}\int_0^t \mathrm{d}s\, H}|\Psi(0)\rangle\,, \tag{6.38}$$

$$A = \int_a^b \mathrm{d}x \, |x\rangle\langle x| \,, \tag{6.39}$$

$$\langle n| = \langle \mathrm{out}_n(t')|\mathrm{e}^{-\mathrm{i}\int_t^{t'} \mathrm{d}s\, H} \,, \tag{6.40}$$

where $|\Psi(0)\rangle$ is the initial state of the particle (far to the left of the barrier) and $\{\langle \mathrm{out}_n(t')| \, |n = \mathrm{t}, \mathrm{r}\} = \{\langle \mathrm{trans}(t')|, \ \langle \mathrm{refl}(t')|\}$ is a set of transmitted and reflected states at a later time t' $(0 < t < t')$. Substituting (6.38)–(6.40) into (6.37) and taking the limit $t' \to \infty$, we obtain (6.36) integrated over x from a to b. Identifying the resultant equation multiplied by $m/\hbar k$ with (6.21), we have[17]

$$\begin{aligned} \tau_\mathrm{t} &= \frac{m}{\hbar k} A_\mathrm{w}^{(\mathrm{t})} = \frac{m}{\hbar k} \frac{\langle \mathrm{trans}(t)|A|\Psi(t)\rangle}{\langle \mathrm{trans}(t)|\Psi(t)\rangle} \\ &= \frac{m}{\hbar k} \frac{1}{\langle \mathrm{trans}(t)|\Psi(t)\rangle} \int_a^b \mathrm{d}x \, \Psi_\mathrm{t}^*(x,t)\Psi(x,t) \,, \end{aligned} \tag{6.41}$$

where $\langle \mathrm{trans}(t)| \equiv \langle \mathrm{trans}(t')|\mathrm{e}^{-\mathrm{i}\int_t^{t'} \mathrm{d}s\, H}$, $\Psi_\mathrm{t}(x,t) \equiv \langle x|\mathrm{trans}(t)\rangle$, and $\Psi(x,t) \equiv \langle x|\Psi(t)\rangle$.[18] We also obtain a similar expression for τ_r. Let us consider a function $\alpha(x)$ that is unity if x is inside the barrier and zero otherwise. Equation (6.41) shows that τ_t can be viewed as the conditional average of $\alpha(x)$ calculated with the conditional probability

$$P(x,t|\mathrm{trans}) = \frac{\Psi_\mathrm{t}^*(x,t)\Psi(x,t)}{\langle \mathrm{trans}(t)|\Psi(t)\rangle} \,. \tag{6.42}$$

Note that this is a spatial average, not a temporal average.

We saw that it is actually possible to construct an equation of the form (6.21) from the probabilistic statement (6.36) with explicit expressions for the conditional probabilities $P(x,t|\mathrm{trans})$ and $P(x,t|\mathrm{trans})$. However, the conditional probabilities and the resultant times (τ_t and τ_r) are in general complex-valued, so that we cannot take those quantities literally. From the viewpoint of dwell time decomposition, the status of the present method is similar to that of the systematic projector approach.

Steinberg [75] has shown that τ_t given by (6.41) agrees with Sokolovski and Baskin's complex tunneling time τ_1 given by (6.24). See also [77] for the connection between the times obtained by the present method and those obtained by other methods. In the weak measurement theory, $A_\mathrm{w}^{(n)}$ is called the *weak value* of A in the time interval between pre-selection of the state $|\Psi(0)\rangle$ and post-selection of the state $|\mathrm{out}_n(t')\rangle$. In the weak measurement theory, the system of interest interacts with an apparatus (measuring device) in a certain manner and what we measure is the position of the pointer of the

[17] The superscript t attached to A_w, which simply means 'transmitted', must not be confused with the lower bound t of the s integration.

[18] $\langle \mathrm{trans}(t)|\Psi(t)\rangle$ is in fact time independent. See [75] for details.

apparatus. A weak value then has the following meaning: when the pointer position is measured, the real part of the weak value gives the mean shift in the pointer position, while the imaginary part gives a shift in the pointer momentum. It thus follows that in the weak measurement theory the real part and the imaginary part of the complex tunneling times are related to such physical shifts in apparatus variables.

6.2.10 Bohm Trajectory Approach

Leavens [46] has shown that it is indeed possible within Bohmian mechanics to find positive definite τ_{t} and τ_{r} that satisfy (6.21). It is a direct consequence of the basic property of Bohm trajectories that they never intersect after starting from different initial positions. In Bohmian mechanics, a quantum mechanical particle has a well-defined position $x(t)$ and velocity $v(t)$ at each time. The velocity field is given by

$$v(x,t) = \left. \frac{J(x,t)}{|\Psi(x,t)|^2} \right|_{x=x(t)} , \tag{6.43}$$

where J is the usual probability current density. Solving the particle's equation of motion $\mathrm{d}x(t)/\mathrm{d}t = v(x(t),t)$ yields a causally determined trajectory $x(x^{(0)},t)$, where $x(0) = x(x^{(0)},0)$ is the particle's initial position. The dwell time for a Bohm trajectory $x(x^{(0)},t)$, the time the trajectory spends in the spatial region of interest,[19] is defined by

$$\tau_{\mathrm{d}}\big(x_1, x_2; x^{(0)}\big) = \int_0^\infty \mathrm{d}t \int_{x_1}^{x_2} \mathrm{d}x\, \delta\big(x - x(x^{(0)},t)\big) . \tag{6.44}$$

Since the initial positions of the trajectories distribute according to the probability distribution $|\Psi(x^{(0)},0)|^2$, the average dwell time is given by

$$\tau_{\mathrm{d}}(x_1, x_2) = \int_{-\infty}^{\infty} \mathrm{d}x^{(0)}\, \tau_{\mathrm{d}}(x_1, x_2; x^{(0)}) |\Psi(x^{(0)},0)|^2 . \tag{6.45}$$

Substituting (6.44) into (6.45) and carrying out the $x^{(0)}$ integral first, we obtain

$$\tau_{\mathrm{d}}(x_1, x_2) = \int_0^\infty \mathrm{d}t \int_{x_1}^{x_2} \mathrm{d}x\, |\Psi(x,t)|^2 . \tag{6.46}$$

The left-hand side of (6.20) is indeed, within Bohmian mechanics, an average dwell time. To decompose $\tau_{\mathrm{d}}(x_1, x_2)$ into transmitted and reflected components, we begin by noting that there exists a special trajectory $x_{\mathrm{c}}(t)$ that separates transmitted and reflected trajectories at any time, as implied by

[19] Let us consider here the dwell time in a general spatial region $x_1 < x < x_2$. The dwell time in the barrier region is immediate by putting $x_1 = a$ and $x_2 = b$.

the non-intersecting property of Bohmian trajectories. The trajectory $x_{\mathrm{c}}(t)$ is defined by

$$\int_{x_{\mathrm{c}}(t)}^{\infty} \mathrm{d}x\, |\Psi(x,t)|^2 = \text{transmission probability } (P_t)\,. \tag{6.47}$$

We can decompose the probability density as

$$|\Psi(x,t)|^2 = \Theta(x - x_{\mathrm{c}}(t))|\Psi(x,t)|^2 + \Theta(x_{\mathrm{c}}(t) - x)|\Psi(x,t)|^2\,. \tag{6.48}$$

Since $x_{\mathrm{c}}(t)$ separates transmitted and reflected trajectories at any time t, the first term on the right-hand side of (6.48) is the to-be-transmitted component of $|\Psi(x,t)|^2$, while the second term is the to-be-reflected component. Integrating both sides of (6.48) with respect to x and t, we immediately obtain

$$\tau_{\mathrm{d}}(x_1,x_2) = P_{\mathrm{t}}\tau_{\mathrm{t}}(x_1,x_2) + P_{\mathrm{r}}\tau_{\mathrm{r}}(x_1,x_2)\,, \tag{6.49}$$

$$\tau_{\mathrm{t}}(x_1,x_2) = \frac{1}{P_{\mathrm{t}}}\int_0^{\infty}\mathrm{d}t\int_{x_1}^{x_2}\mathrm{d}x\,\Theta(x - x_{\mathrm{c}}(t))|\Psi(x,t)|^2\,, \tag{6.50}$$

$$\tau_{\mathrm{r}}(x_1,x_2) = \frac{1}{P_{\mathrm{r}}}\int_0^{\infty}\mathrm{d}t\int_{x_1}^{x_2}\mathrm{d}x\,\Theta(x_{\mathrm{c}}(t) - x)|\Psi(x,t)|^2\,. \tag{6.51}$$

Equation (6.49) has the same form as (6.21) with positive definite times on its right-hand side. Moreover, τ_{d}, τ_{t}, and τ_{r} above satisfy the additivity condition (6.30). For now, the use of Bohmian mechanics is the only way to arrive at this remarkable result. Leavens, at the same time, carefully points out the price one has to pay. That is, the connection between the case of a wave packet and the case of a stationary state is not simple:

$$\tau_{\mathrm{t}}(x_1,x_2) \neq \frac{1}{P_t}\int\frac{\mathrm{d}k}{2\pi}|\varphi(k)|^2|T(k)|^2\tau_{\mathrm{t}}(x_1,x_2;k)\,, \tag{6.52}$$

where $\tau_{\mathrm{t}}(x_1,x_2;k)$ is the transmission time for a stationary case, and similarly reflection time. The lack of the simple relation is due to the fact that $\tau_{\mathrm{t}}(x_1,x_2)$ and $\tau_{\mathrm{r}}(x_1,x_2)$ derived above are not bilinear in $\Psi(x,t)$, because they depend on $x_{\mathrm{c}}(t)$, and $x_{\mathrm{c}}(t)$ depends on $\Psi(x,t)$ through (6.47).

The separability of Bohm trajectories into to-be-transmitted and to-be-reflected parts is at the heart of the derivation of (6.49)–(6.51). This property is often criticised as unphysical. However, the possibility remains that the results obtained are correct and justifiable even if the separation itself is unphysical. At the same time, the status of (6.49)–(6.51) in conventional quantum mechanics needs to be clarified. More fundamentally, we must keep in mind, before arguing the physical meaning of (6.49)–(6.51), that the decomposition (6.21) itself could be meaningless in quantum mechanics. We should not take (6.21) for granted, simply because our classical intuition behind it could be wrong for purely quantum mechanical phenomena.

For Leavens' earlier contributions regarding Bohm trajectories, see [80,81]. One of the conclusions is that a Bohm-trajectory calculation of tunneling time gives answers that are very different from those obtained from the Larmor clock or the time-modulated barrier method by Büttiker and Landauer.

6.3 Tunneling Times as Interfering Alternatives: A Feynman Paths Approach

6.3.1 Lesson from Two-Slit Experiments

I claimed in [9] that we should understand the nature of tunneling time from the viewpoint of *interfering alternatives* and argued that what makes sense in the tunneling time problem is the *range* of tunneling times. An important point is that probabilities cannot be considered in the range. We cannot say that the tunneling time takes a certain value in the range with such and such a probability. This is because different values of tunneling time are interfering alternatives, i.e., they are not exclusive. Recall that probabilities are defined only for an exhaustive set of mutually exclusive alternatives. An analogy with the two-slit experiment would be helpful to understand this point. In the two-slit experiment, due to the interference between the two alternatives of a particle going through the upper slit and going through the lower slit, probabilities cannot be defined for the two alternatives without disturbing the motion of the particle (see Fig. 6.7).

In the tunneling time problem, the y axis (at $x = 0$) on which we considered position alternatives corresponds to the time axis on which we consider tunneling time alternatives (alternatives correspond to different values of tunneling time). To discuss quantum interference between tunneling time alternatives, we need a measure of the interference. We can find it by working in the real time formulation of Feynman path integrals. The starting

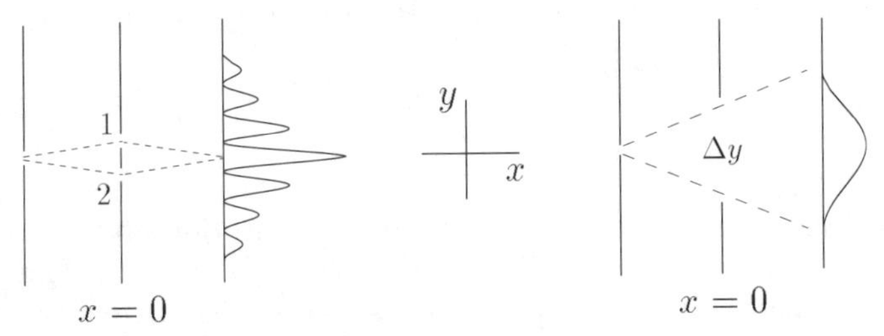

Fig. 6.7. *Left*: we cannot say that 'a particle passes through slit 1 or slit 2 with some probability' in the two-slit experiment. *Right*: we can say, however, that 'a particle passes through the slit of width Δy with probability approximately unity' in the one-slit experiment. Therefore, although the precise point the particle passes through cannot be considered, the range of space (Δy) it passes through can be considered

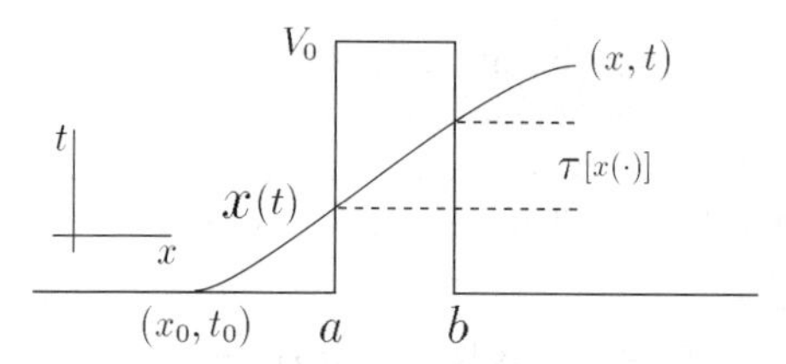

Fig. 6.8. The time a Feynman path spends in the barrier

point is the amount of time $\tau[x(\cdot)]$ which a Feynman path $x(\cdot)$ traversing the barrier spends inside the barrier region. Since Feynman paths can cross the two points $x = a$ and $x = b$ many times (in Fig. 6.8, the single crossing case is shown), several definitions are possible for $\tau[x(\cdot)]$. In [9], I used the 'resident-time' definition widely used in the literature:[20]

$$\tau[x(\cdot)] \equiv \int_{t_0}^{t} \mathrm{d}t' \Theta_{ab}(x(t')) , \tag{6.53}$$

where $\Theta_{ab}(x)$ is unity for $a < x < b$ and zero otherwise.

Since there is no classical solution to the equation of motion in the tunneling regime, many paths contribute to the tunneling propagator and the value of $\tau[x(\cdot)]$ is not unique. We can classify the paths according to the values of $\tau[x(\cdot)]$. The sum over paths defining the tunneling propagator can then be written as a time integral over all values of $\tau[x(\cdot)]$, and as a result we can write the transmitted packet as

$$\Psi(x,t) = \int \mathrm{d}\tau \, \Psi(x,t|\tau) , \tag{6.54}$$

where $\Psi(x,t|\tau)$ is the product of the 'partial propagator' $K(x,t;x_0,t_0|\tau)$ to which only the paths satisfying $\tau[x(\cdot)] = \tau$ contribute and the initial state $\Psi(x_0,t_0)$ that is nonvanishing only on the left-hand side of the barrier, followed by an x_0 integral. What corresponds to (6.54) in the two-slit experiment is

$$\Psi(y,t) = \sum_{j=1,2} \Psi_j(y,t) , \tag{6.55}$$

where $\Psi(y,t)$ (omitting x dependence) is the total wave function after the slits and $\Psi_j(y,t)$ is the partial wave function from slit j ($j = 1,2$). The probability that a particle is found at point y on the final screen at time t is given by $|\Psi(y,t)|^2 = |\Psi_1(y,t) + \Psi_2(y,t)|^2$, where x implicit in Ψ is supposed to correspond to the screen position. The alternative that a particle goes

[20] In [82] I used a 'passage-time' definition and applied the CHA criterion to the tunneling time problem for the first time. The passage-time definition was first employed by Schulman and Ziolkowski [52].

through slit 1 and arrives at y and another alternative that it goes through slit 2 and arrives at y are not exclusive because

$$|\Psi_1(y,t) + \Psi_2(y,t)|^2 \neq |\Psi_1(y,t)|^2 + |\Psi_2(y,t)|^2 . \tag{6.56}$$

The cross term

$$\mathrm{Re}\, \Psi_1^*(y,t)\Psi_2(y,t) \tag{6.57}$$

is the measure of interference between the two alternatives.

6.3.2 Tunneling Time Probability Distribution is not Definable

Now, for the tunneling time problem, we follow the same line as above. The tunneling probability, being the probability that a particle is found somewhere on the other side of the barrier at sufficiently later times, is given by

$$P \equiv \lim_{t\to\infty} \int_b^\infty \mathrm{d}x\, |\Psi(x,t)|^2 . \tag{6.58}$$

Substituting (6.54) into (6.58), we can write

$$P = \int \mathrm{d}\tau \int \mathrm{d}\tau'\, D[\tau;\tau'] , \tag{6.59}$$

where

$$D[\tau;\tau'] \equiv \lim_{t\to\infty} \int_b^\infty \mathrm{d}x\ \Psi^*(x,t|\tau)\Psi(x,t|\tau') . \tag{6.60}$$

Note that only the real part of D contributes to the double τ integrals since $D^*[\tau;\tau'] = D[\tau';\tau]$. Now, Re$D[\tau;\tau']$ gives the measure of interference between the two alternatives of a particle spending time τ and spending time τ' in the barrier region. It is, of course, a subject for investigation whether the interference vanishes or not, i.e., whether an equation of the form

$$\mathrm{Re}\, D[\tau;\tau'] = \delta(\tau-\tau')P(\tau) \tag{6.61}$$

holds or not. Equation (6.61) reads as follows: $\mathrm{Re}\, D[\tau;\tau']$ is proportional to $\delta(\tau-\tau')$ with a proportionality coefficient $P(\tau)$. To clarify the physical meaning of $P(\tau)$, let us see what happens if we actually assume (6.61). Under this assumption, one finds

$$P(\tau) \geq 0 \quad \text{and} \quad \int \mathrm{d}\tau\, P(\tau) = P , \tag{6.62}$$

which are exactly the properties that a probability distribution of tunneling times must have. Based on this observation, I consider (6.61) to be the criterion to judge the possibility of a probabilistic description of tunneling time: *If (6.61) holds, a probability distribution of tunneling time is definable* *and it is given by the proportionality coefficient $P(\tau)$; otherwise, a probability distribution is not definable.*

To investigate whether (6.61) holds or not, we must calculate the right-hand side of (6.60) using a concrete expression for $\Psi(x,t|\tau)$. This was done in [9] using (3.3) in [66], and I shall only repeat the consequence here. $D[\tau;\tau']$ was found to be

$$D[\tau;\tau'] = \mathrm{e}^{\mathrm{i}V_0(\tau-\tau')/\hbar}\langle \mathcal{T}_k^*(\tau)\mathcal{T}_k(\tau')\rangle\,, \tag{6.63}$$

where

$$\mathcal{T}_k(\tau) \equiv \int \frac{\mathrm{d}V}{2\pi\hbar}\mathrm{e}^{\mathrm{i}V\tau/\hbar}T(V,k)\,, \tag{6.64}$$

and $\langle\cdots\rangle \equiv \int \mathrm{d}k \cdots |\psi(k)|^2$ with $\psi(k)$ the k-space wave function of the incident particle and $T(V,k)$ the transmission amplitude when the potential height is V. The range of the V integration is $(-\infty,\infty)$. To guarantee that the incident particle moves towards the barrier with energy less than the barrier height, $\psi(k)$ is so restricted that it vanishes for $k<0$ and for $k>\sqrt{2mV_0}/\hbar$, and satisfies the normalization $\int \mathrm{d}k\,|\psi(k)|^2 = 1$. I found that (6.61) with (6.63) leads to

$$P_{V_0-V} + P_{V_0+V} = 2P_{V_0}\,, \tag{6.65}$$

for arbitrary V and then showed that (6.65) leads to a contradiction, namely that the tunneling probability is higher for higher barriers. This means that (6.61) does not hold. The physical conclusion is, therefore, that a probability distribution for tunneling times is not definable since different values of τ interfere.

6.3.3 Relation to Gell-Mann and Hartle's Formulation of Quantum Mechanics

In terms of Gell-Mann and Hartle's generalized quantum mechanics (GQM) (or, the decoherent histories approach to quantum mechanics) [13,14], $D[\tau;\tau']$ and (6.61) are, respectively, the decoherence functional and the weak decoherence condition (WDC) [83]. GQM provides a general scheme for the study of quantum probabilities for 'histories', which are alternatives not restricted to a moment of time. The negative result reached above may be regarded as a result of GQM with continuous histories (Feynman paths), rather than discrete histories (the products of projection operators at different times). Although GQM with continuous histories has received less attention, our study demonstrates how it can be applied to concrete problems.

6.3.4 $\tau_\mathrm{d} = P_\mathrm{t}\tau_\mathrm{t} + P_\mathrm{r}\tau_\mathrm{r}$ Is Doubtful

In situations where discussions on the tunneling time problem do not converge, some authors proposed criteria for 'correct' tunneling times. The most widely used among them is the following 'probabilistic identity' introduced by Hauge and Støvneng in [2]:

$$\tau_{\rm d} = P_{\rm t}\tau_{\rm t} + P_{\rm r}\tau_{\rm r}\,, \tag{6.66}$$

where $\tau_{\rm d}$ is the dwell time, the total average time a particle spends in the barrier region irrespective of whether it is finally reflected or transmitted, $\tau_{\rm t}$ is the transmission time, the average time a transmitted particle spends in the barrier, and $\tau_{\rm r}$ is the reflection time, the average time a reflected particle spends in the barrier. $P_{\rm t}$ is the transmission probability and $P_{\rm r}$ the reflection probability, satisfying $P_{\rm t} + P_{\rm r} = 1$. Here, as well as in [9], we concentrate on the transmission time, calling it the tunneling time (our P is $P_{\rm t}$). In many arguments, $\tau_{\rm d}$, $\tau_{\rm t}$, and $\tau_{\rm r}$ in Eq. (6.66) are assumed (often implicitly) to be averaged quantities. In other words, the existence of the probability distributions of these times are often assumed behind Eq. (6.66). In fact, Eq. (6.66) seems most rigorously founded if such distributions exist. As concluded above, however, the probability distribution is not definable, at least, for $\tau_{\rm t}$. (It is also unlikely that $\tau_{\rm r}$ and $\tau_{\rm d}$ allow probabilistic descriptions because WDC is a very severe condition.) Thus, the validity of Eq. (6.66) is doubtful. In my opinion, many arguments in which Eq. (6.66) plays essential roles should be reconsidered.

At the same time, it should be pointed out that our negative result does not exclude the possibility of a distribution of τ that is not fully inherent in tunneling but is valid only with a certain measurement setup. Although such non-intrinsic distributions with limited validity are not the subject here, the probabilistic identity could find its meaning with such non-intrinsic distributions. Sokolovski and coworkers [84,85] have proposed distributions of tunneling times by using a Larmor clock as the measuring device, where the angle of spin rotation due to a magnetic field in the barrier is measured and compared with theory.

6.3.5 Range of Tunneling Times

From a classical point of view, it is hard to understand in the two-slit experiment that going through the upper slit and going through the lower slit are not exclusive, but this is how nature behaves and we just accept it. In the same way, we must accept the non-exclusiveness of tunneling times and give up a probabilistic description of tunneling time. This should not be regarded as a 'discouraging' result. It just shows how nature behaves. What we need to do is to consider what is meaningful when nature behaves that way. An idea arises from the one-slit experiment (see Fig. 6.7): just as the spatial range Δy the particle goes through can be meaningfully considered, we should be able to consider the range of tunneling times. Let us denote by $\mathcal{C}_\tau$ the class of Feynman paths for which $\tau[x(\cdot)] = \tau$. Roughly speaking, the range would be such that the sum over paths for the tunneling process is approximated

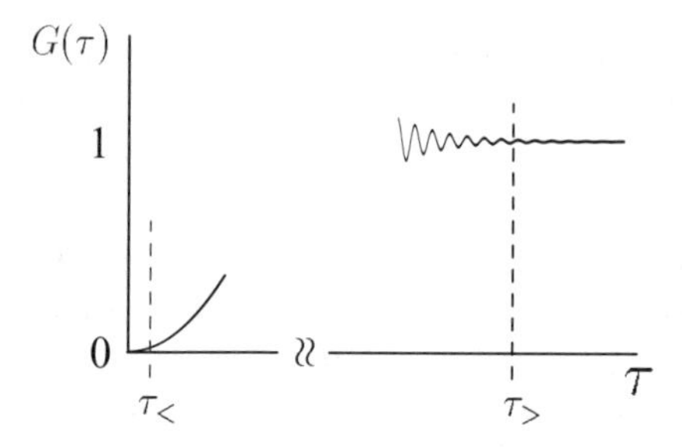

Fig. 6.9. How to estimate the range of tunneling times using the function $G(\tau)$

with sufficient accuracy by the sum over those paths that belong to $\mathcal{C}_\tau$ with τ exhausting all values in the range. More precisely, when

$$\int_s^t \mathrm{d}\tau \int_s^t \mathrm{d}\tau' \, D[\tau;\tau'] \approx P \tag{6.67}$$

holds with sufficient accuracy for $\forall s < \tau_<$ and for $\forall t > \tau_>$, I propose to regard $(\tau_<, \tau_>)$ as the range of tunneling times. Rather than dealing with $\int_s^t \mathrm{d}\tau \int_s^t \mathrm{d}\tau' \, D[\tau;\tau']$, which is a function of two variables s and t, let us work with a function of one variable

$$G(\tau) \equiv \frac{1}{P} \int_0^\tau \mathrm{d}\tau' \int_0^\tau \mathrm{d}\tau'' \, D[\tau';\tau''] \,, \tag{6.68}$$

which satisfies $G(0) = 0$ and $G(\infty) = 1$. The range of tunneling times can then be estimated by finding $\tau_<$ and $\tau_>$ such that, for $\epsilon \ll 1$,

$$G(\tau) < \epsilon \ \text{for} \ \forall \tau < \tau_<; \quad |1 - G(\tau)| < \epsilon \ \text{for} \ \forall \tau > \tau_> \,. \tag{6.69}$$

See Fig. 6.9).

It is interesting to see that, if WDC [(6.61)] held (though it does not in fact hold), the range of tunneling times would simply be the support of the probability distribution of tunneling time. This is because, assuming (6.61), (6.67) reduces to

$$\int_s^t \mathrm{d}\tau \, P(\tau) \approx P \,. \tag{6.70}$$

It would be natural to call the support (s, t) of $P(\tau)$ the range of tunneling times if $P(\tau)$ were definable. The concept of the range of tunneling times can be introduced, however, regardless of the existence of $P(\tau)$. In (6.67), the range of tunneling times is defined at the level of amplitudes, not at the level of probabilities.

The function $G(\tau)$ can only be obtained analytically for delta function potentials, so we must rely on numerical calculations. From (6.63) and (6.68),

$$G(\tau) = \frac{1}{P} \int \mathrm{d}k \, |\psi(k)|^2 \left| \int_0^\tau \mathrm{d}\tau' \, \mathrm{e}^{-\mathrm{i}V_0\tau'/\hbar} \int_{-\infty}^{\infty} \frac{\mathrm{d}V}{2\pi\hbar} \mathrm{e}^{\mathrm{i}V\tau'/\hbar} T(V,k) \right|^2 , \tag{6.71}$$

$$P = \int \mathrm{d}k \; |\psi(k)|^2 \, |T(V_0, k)|^2 \; . \tag{6.72}$$

It is clear that the initial state of the incident particle [or, $\psi(k)$] and the barrier shape [or, $T(V, k)$] determine the function $G(\tau)$ and thus determine the range of tunneling times. More fundamentally, the decoherence functional $D[\tau; \tau']$ is written in terms of the initial state and the barrier shape [see (6.63)]. It is quite reasonable that the two elements (the particle and the barrier) finally determine speakables and unspeakables in the tunneling time problem. Equation (6.71) is not only of conceptual importance but also of practical importance as a basis for numerical calculations of $G(\tau)$.

6.3.6 An Example: The Monochromatic Case

Here, following [9], let us consider the case of an incident particle having a constant energy. We thus put $|\psi(k)|^2 = \delta(k - k_0)$, which leads to

$$G(\tau) = \left| \frac{1}{T(V_0, k_0)} \int_{infty}^{\infty} \frac{\mathrm{d}V}{\pi} \, \frac{\sin(V\tau/\hbar)}{V} \, T(V_0 + V, k_0) \right|^2 \; , \tag{6.73}$$

where $T(V, k)$ is given by (6.4) with repacement $V_0 \to V$. Inside the square is the function $C(\tau)$ introduced by Fertig [86,87]. To be precise, our formulation loses its validity when the incident energy is constant because the assumption of well-defined reflected and transmitted packets, which is used in the derivation of (6.63), is not satisfied in such a case. Therefore, $|\psi(k)|^2 = \delta(k - k_0)$ should be understood as $|\psi(k)|^2 \to \delta(k - k_0)$. Figure 6.10 shows a typical example of $G(\tau)$ for opaque barriers in the monochromatic limit, where the free passage time of the region, $\tau_{\mathrm{f}} \equiv md/\hbar k_0$, is used as the time unit. The parameter values are $k_0 d = 1$ ($d \equiv b - a$) and $V_0/E_0 = 5$ [$E_0 \equiv (\hbar k_0)^2/2m$], for which $P = |T(V_0, k_0)|^2 = 4.64 \times 10^{-2}$. From Fig. 6.10, we can read, by choosing $\epsilon = 0.01$ for instance, $\tau_{<}/\tau_{\mathrm{f}} \approx 0.0335$ and $\tau_{>}/\tau_{\mathrm{f}} \approx 5.95$.

Among various tunneling times proposed in the literature, the most famous ones are the Larmor time $\tau_{\mathrm{LM}} = \hbar/V_0\sqrt{V_0/E_0 - 1}$ [29,30] and the Büttiker–Landauer time $\tau_{\mathrm{BL}} = d\sqrt{m/2(V_0 - E_0)}$ [34]. (Another famous one is the phase time.) In Fig. 6.10, LM and BL indicate the Larmor time and the Büttiker–Landauer time, respectively. Note that $\tau_{<} \ll \tau_{\mathrm{LM}}$ and $\tau_{>} \gg \tau_{\mathrm{BL}}$. This is true for a wide range of parameters corresponding to opaque barriers. If both $\tau_{<}$ and $\tau_{>}$ were close to τ_{BL} (or τ_{LM}), then we would be able to regard τ_{BL} (or τ_{LM}) as 'the tunneling time' in a loose sense, but this is not the case. In [52,65], τ_{BL} was considered as a measure of the range of tunneling times, but the range obtained here is much wider than τ_{BL}.

In addition to $\tau_{<}$ and $\tau_{>}$, we can find (though only loosely) another characteristic time τ_{c}. This is the time at which $G(\tau)$ takes its maximum when its noisy behavior is ignored. Figure 6.10 shows $\tau_{\mathrm{c}} \sim \tau_{\mathrm{BL}}$, which is also widely observed in our numerical results for opaque barriers. Although $G(\tau)$ takes significantly large values around τ_{c}, we cannot regard $\mathcal{C}_{\tau_{\mathrm{c}}}$ as the dominantly

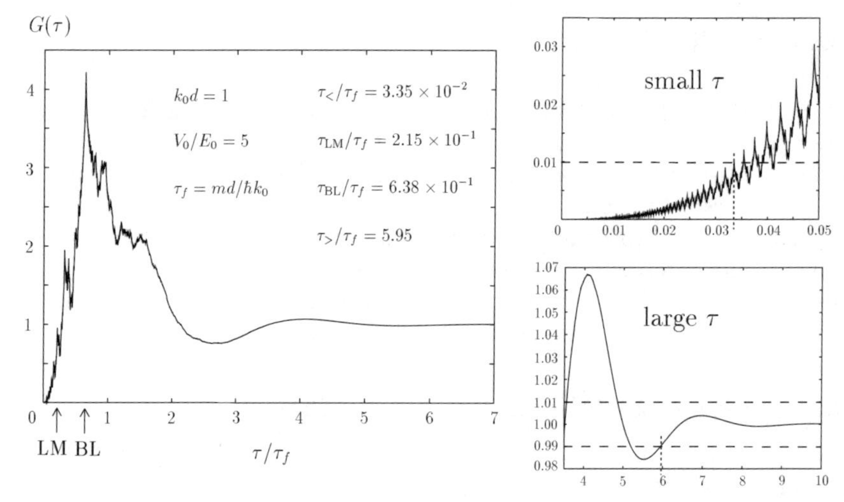

Fig. 6.10. $G(\tau)$ for the monochromatic case

contributing class of paths, since, as implied by the decreasing behavior of $G(\tau)$ for larger τ, the contribution from the class is mostly canceled by contributions from other classes of paths. The idea of identifying 'the tunneling time' by finding a sharply defined 'dominantly contributing class of paths' does not work. There is no such class.

We have found three characteristic times associated with tunneling: $\tau_<$, $\tau_>$, and τ_c. The physical meaning of these times would be clarified in situations where the particle interacts with other degrees of freedom in the barrier. I would expect that $\tau_<^{-1}$ and $\tau_>^{-1}$ determine the parts of the environment (in terms of its frequencies) that can affect the motion of the tunneling particle. The large values of $G(\tau)$ around $\tau = \tau_c$ might be an implication that the tunneling probability would be greatly affected if the Feynman paths contributions from around $\mathcal{C}_{\tau_c}$ are altered. If so, a time-dependent interaction whose time scale τ_{int} is about τ_c would cause a significant change in the tunneling probability. The significant change would occur not at a specific value of τ_{int} but over a range of τ_{int} around $\tau_{\text{int}} = \tau_c$ because $G(\tau)$ takes large values *around* $\tau = \tau_c$.

The width of the range of tunneling times is essentially determined by $\tau_>$. In [9], I derived an asymptotic expression for $G(\tau)$ to give a formula for $\tau_>$ for opaque barriers:

$$\tau_> \sim \frac{\hbar}{\text{Im}(Z_0)}\left[l\sqrt{u-1} + \ln\left|\frac{r\sqrt{u}}{2\epsilon(z_0-u)}\right|\right], \quad u \equiv \frac{V_0}{E_0}, \; l \equiv k_0 d, \tag{6.74}$$

where $z_0 \equiv Z_0/E_0$, $r \equiv R/E_0 = 4(1-z_0)/(2-\mathrm{i}lz_0)$. Among the poles of $T(V,k)$ in the complex V plane, Z_0 is the pole that is closest to the real V axis, and R is the residue of $T(V,k)$ at the pole. A graphical method for obtaining z_0 is given in [9].

6.4 Conclusion

In my opinion, the main reason for the long-lasting controversy over the tunneling time problem is that it has not been noticed[21] that different values of tunneling time are not exclusive. Just as probabilities cannot be considered for the non-exclusive alternative of which slit a particle passes through in the two-slit experiment, probabilities cannot be considered for different values of tunneling time since they are not exclusive. What can be meaningfully defined is the range of space the particle passes through in the one-slit experiment and the range of tunneling times in the tunneling time problem. The range of tunneling times is so broad for opaque barriers that a unique tunneling time cannot be considered even in a loose sense.

A significant of the range of tunneling times is suggested even from the preliminary investigations of Sect. 6.3; various tunneling times that were proposed in the past are found to fall within the range, implying that the information of important time scales of tunneling is 'embedded' in the range. Moreover, I expect that the further studies of the range in the presence of other degrees of freedom will provide a new insight into how the coupling between the particle and the environments affects the dynamics of tunneling and hence alters the tunneling probabilities or the tunneling currents. Through these studies, we would be able to learn how the tunneling time range can be measured experimentally and how the system information can be extracted from the range. The range of tunneling times under the influence of environments thus seems to be an exciting direction to explore toward the 'tunneling time spectroscopy'.

Acknowledgments

I thank Dr. S. Takagi, Dr. C.R. Leavens, Dr. N. García, and Dr. Nejo for stimulating discussions.

[21] Exceptions are [84] (see Sect. X there) and [87], in which completely different approaches from our own were pursued for probability distributions of tunneling time, although they also worked with real-time Feynman path integrals. A fundamental difference between their approaches and the present one is that we have a solid criterion (WDC) for judging the definability of a probability distribution. Definability thus becomes a subject for investigation, rather than an assumption. Another difference is that the present approach introduces the concept of the range of tunneling times in a precise manner based on the decoherence functional.

References

1. R. Landauer and T. Martin: Rev. Mod. Phys. **66**, 217 (1994)
2. E.H. Hauge and J.A. Støvneng: Rev. Mod. Phys. **61**, 917 (1989)
3. V.S. Olkhovsky and E. Recami: Phys. Rep. **214**, 339 (1992)
4. M. Büttiker: In: *Electronic Properties of Multilayers and low Dimensional Semiconductors*, ed. by J.M. Chamberlain, L. Eaves, and J.C. Portal, (Plenum, New York 1990) p. 297
5. C.R. Leavens and G.C. Aers: In: *Scanning Tunneling Microscopy and Related Methods*, ed. by R.J. Behm, N. Garcia and H. Rohrer (Kluwer, Dordrecht 1990) p. 59
6. M. Jonson: In: *Quantum Transport in Semiconductors*, ed. by D.K. Ferry and C. Jacoboni (Plenum, New York 1991) p. 193
7. A.P. Jauho: In: *Hot Carriers in Semiconductor Nanostructures: Physics and Applications*, ed. by J. Shah (Academic, Boston 1992) p. 121
8. *Proceedings of the Adriatico Research Conference on Tunneling and its Implications*, ed. by D. Mugnai, A. Ranfagni, L.S. Schulman (World Scientific, Singapore 1997)
9. N. Yamada: Phys. Rev. Lett. **83**, 3350 (1999)
10. R.B. Griffiths: J. Stat. Phys. **36**, 219 (1984)
11. R. Omnès: J. Stat. Phys. **53**, 893 (1988)
12. R. Omnès: *The Interpretation of Quantum Mechanics* (Princeton University Press, Princeton, NJ 1994)
13. M. Gell-Mann and J.B. Hartle: In: *Proceedings of the 3rd International Symposium on the Foundations of Quantum Mechanics in the Light of New Technology*, ed. by S. Kobayashi, H. Ezawa, Y. Murayama, and S. Nomura (Physical Society of Japan, Tokyo 1990)
14. J.B. Hartle: Phys. Rev. D **44**, 3173 (1991)
15. N. Yamada and S. Takagi: Prog. Theor. Phys. **85**, 985 (1991)
16. N. Yamada and S. Takagi: Prog. Theor. Phys. **86**, 599 (1991)
17. J.A. Støvneng: *Can One Speak about Tunneling Times in Polite Society?* in [2], pp. 1–17 (see Sect. 4.3)
18. K.W.H. Stevens: J. Phys. C: Solid State Phys. **16**, 3649 (1983)
19. N. Teranishi, A.M. Kriman, and D.K. Ferry: Superlatt. Microstruc. **3**, 509 (1987)
20. A.P. Jauho and M. Jonson: Superlatt. Microstruc. **6**, 303 (1989)
21. A. Ranfagni, D. Mugnai, and A. Agresti: Phys. Lett. A **158**, 161 (1991)
22. P. Moretti: Phys. Rev. A **46**, 1233 (1992)
23. J.G. Muga, S. Brouard, V. Delgado, and J.P. Palao: Characteristic Times in One-Dimensional Tunneling. In: *Proceedings of the Adriatico Research Conference on Tunneling and its Implications*, ed. by D. Mugnai, A. Ranfagni, L.S. Schulman (World Scientific, Singapore, 1997) pp. 34–49
24. S. Brouard and J.G. Muga: Phys. Rev. A **54**, 3055 (1996)
25. G. Garcia-Calderon: Tunneling dynamics of the Plane Wave Shutter. In: *Proceedings of the Adriatico Research Conference on Tunneling and its Implications*, ed. by D. Mugnai, A. Ranfagni, L.S. Schulman (World Scientific, Singapore, 1997) pp. 176–190
26. G. García-Calderón and A. Rubio: Phys. Rev. A **55**, 3361 (1997)
27. A. Goldberg, H.M. Schey, and J.L. Schwartz: Am. J. Phys. **35**, 117 (1967)

28. M. Büttiker: Phys. Rev. B **27**, 6178 (1983)
29. A.I. Baz': Yad. Fiz. **4**, 252 (1966) [Sov. J. Nucl. Phys. **4**, 182 (1967)]
30. V.F. Rybachenko: Yad. Fiz. **5**, 895 (1966) [Sov. J. Nucl. Phys. **5**, 635 (1967)]
31. C.R. Leavens and G.C. Aers: Solid State Commun. **63**, 1101 (1987)
32. C.R. Leavens and G.C. Aers: Solid State Commun. **67**, 1135 (1988)
33. M. Hino, N. Achiwa, S. Tasaki, T. Ebisawa, T. Kawai, T. Akiyoshi, and D. Yamazaki: Phys. Rev. A **59**, 2261 (1999)
34. M. Büttiker and R. Landauer: Phys. Rev. Lett. **49**, 1739 (1982)
35. M. Büttiker and R. Landauer: Phys. Scr. **32**, 429 (1985)
36. S. Takagi: Tunneling Through a Squeezing Barrier: A Comment on the Barrier Traversal Time. In: *Proceedings of the 4th International Symposium on the Foundations of Quantum Mechanics in the Light of New Technology*, Tokyo, August 24–27, 1992, ed. by M. Tsukada et al. (JJAP, Tokyo 1993) pp. 82–85
37. W.S. Truscott: Phys. Rev. A **70**, 1900 (1993)
38. A.P. Jauho and M. Jonson: J. Phys. Condens. Matter **1**, 9027 (1989)
39. H. De Raedt, N. García, and J. Huyghebaert: Solid State Commun. **76**, 847 (1990)
40. C.R. Leavens and G.C. Aers: Solid State Commun. **78**, 1015 (1991)
41. F.T. Smith: Phys. Rev. **118**, 349 (1960)
42. C.R. Leavens and G.C. Aers: Phys. Rev. B **39**, 1202 (1989)
43. E.H. Hauge, J.P. Falck, and T.A. Fjeldly: Phy. Rev. B **36**, 4203 (1987)
44. L.M. Baskin and D.G. Sokolovski: Sov. Phys. J. **30**, 204 (1987)
45. D. Sokolovski and L.M. Baskin: Phys. Rev. A **36**, 4604 (1987)
46. C.R. Leavens and G.C. Aers: Bohm Trajectories and the Tunneling Time Problem. In: *Scanning Tunneling Microscopy III*, ed. by R. Wiesendanger and H.-J. Güntherodt (Springer, Berlin, Heidelberg, New York 1993) pp. 105–140
47. W. Jaworski and D. Wardlaw: Phys. Rev. A **37** 2843 (1988)
48. W. Jaworski and D. Wardlaw: Phys. Rev. A **40** 6210 (1989)
49. H. Yamamoto, Y. Senshu, K. Miyamoto, and S. Tanaka: Physics. Status Solidi B **206**, 601 (1998)
50. G. Iannaccone and B. Pellegrini: Phys. Rev. B **49**, 16548 (1994)
51. T. Martin: Int. J. Mod. Phys. B **10**, 3747 (1996)
52. L.S. Schulman and R.W. Ziolkowski: In: *Proceedings of Third International Conference on Path Integrals from meV to MeV*, ed. by V. Sa-yakanit et al. (World Scientific, Singapore 1989)
53. A.B. Nassar: Phys. Rev. A **38**, 683 (1988)
54. M.-Q. Chen and M.S. Wang: Phys. Lett. A **149**, 441 (1990)
55. K. Imafuku, I. Ohba, Y. Yamanaka: Phys. Lett. A **204**, 329 (1995)
56. K.L. Jensen and F.A. Buot: Appl. Phys. Lett. **55**, 669 (1989)
57. J.G. Muga, S. Brouard, and R. Sala: Phys. Lett. A **167**, 24 (1992)
58. T.P. Spiller, T.D. Clark, R.J. Prance, and H. Prance: Europhys. Lett. **12**, 1 (1990)
59. E. Pollak and W.H. Miller: Phys. Rev. Lett **53**, 115 (1984)
60. T. Martin and R. Landauer: Phys. Rev. A **47**, 2023 (1993)
61. C.R. Leavens: Superlattices and Microstructures **23**, 795 (1998)
62. C.R. Leavens and G.C. Aers: Phys. Rev. B **40**, 5387 (1989)
63. C.R. Leavens and G.C. Aers: J. Vac. Sci. Technol. A **6**, 305 (1988)
64. D. Sokolovski and J.N.L. Connor: Phys. Rev. A **44**, 1500 (1991)
65. D. Sokolovski and J.N.L. Connor: Solid State Commun. **89**, 475 (1994)

66. D. Sokolovski: Phys. Rev. A **52**, R5 (1995)
67. D. Sokolovski: Phys. Rev. Lett. **79**, 4946 (1997)
68. D. Sokolovski: Phys. Rev. A **57**, R1469 (1998)
69. C.R. Leavens: Found. Phys. **25**, 229 (1995)
70. S. Brouard, R. Sala, and J.G. Muga: Phys. Rev. A **49**, 4312 (1994)
71. J.G. Muga, S. Brouard, and R. Sala: J. Phys. Condens. Matter **4**, L579 (1992)
72. S. Brouard, R. Sala, and J.G. Muga: Europhys. Lett. **22**, 159 (1993)
73. B.A. van Tiggelen, A. Tip, and A. Lagendijk: J. Phys. A **26**, 1731 (1993)
74. C.R. Leavens: Bohmian Mechanics and the Tunneling Time Problem for Electrons. In: *Proceedings of the Adriatico Research Conference on Tunneling and its Implications*, ed. by D. Mugnai, A. Ranfagni, L.S. Schulman (World Scientific, Singapore 1997) pp. 100–120
75. A.M. Steinberg: Phys. Rev. Lett. **74**, 2405 (1995)
76. A.M. Steinberg: Phys. Rev. A **52**, 32 (1995)
77. G. Iannaccone: Weak Measurement and the Traversal Time Problem. In: *Proceedings of the Adriatico Research Conference on Tunneling and its Implications*, ed. by D. Mugnai, A. Ranfagni, L.S. Schulman (World Scientific, Singapore 1997) pp. 292–309
78. Y. Aharonov, D.Z. Albert, and L. Vaidman: Phys. Rev. Lett. **60**, 1351 (1988)
79. Y. Aharonov and L. Vaidman: Phys. Rev. A **41**, 11 (1990)
80. C.R. Leavens: Solid State Commun. **74**, 923 (1990)
81. C.R. Leavens: Solid State Commun. **76**, 253 (1990)
82. N. Yamada: Phys. Rev. A **54**, 182, (1996)
83. M. Gell-Mann and J.B. Hartle: In: *Proceedings of the 25th International Conference on High Energy Physics*, ed. by K.K. Phua and Y. Yamaguchi (World Scientific, Singapore 1991)
84. D. Sokolovski and J.N.L. Connor: Phys. Rev. A **47**, 4677 (1993)
85. D. Sokolovski, S. Brouard, and J.N.L. Connor: Phys. Rev. A **50**, 1240 (1994)
86. H.A. Fertig: Phys. Rev. Lett. **65**, 2321 (1990)
87. H.A. Fertig: Phys. Rev. B **47**, 1346 (1993)
88. D. Sokolovski and J.N.L. Connor: Phys. Rev. A **42**, 6512 (1990)

Index

Printing (Computer to Film): Saladruck Berlin
Binding: Stürtz AG, Würzburg